C000065332

# THE
# ARIZONA
# RANGERS

## Bill O'Neal

EAKIN PRESS ⬦ Fort Worth, Texas

Copyright © 1987
By Bill O'Neal
Published By Eakin Press
An Imprint of Wild Horse Media Group
P.O. Box 331779
Fort Worth, Texas 76163
1-817-344-7036
**www.EakinPress.com**
**ALL RIGHTS RESERVED**
1  2  3  4  5  6  7  8  9
ISBN-10: 0-89015-610-7
ISBN-13: 978-0-89015-610-0

**Library of Congress Cataloging-in-Publication Data**

O'Neal, Bill, 1942-
  The Arizona Rangers. Special Collectors' Edition.

  Bibliography: p.
  Includes index.
  1. Arizona Rangers — History.  2. Frontier and pioneer life — Arizona.
3. Arizona — History.  I. Title.
F811.O48 1987 979.1'04 87-3552
ISBN 0-89015-610-7

*For Lynn, Shellie, Berri, and Causby*
*My own squad of Rangerettes*

# Contents

# Acknowledgments

My first and deepest debt of gratitude is to Lori Davisson, research specialist of the Arizona Heritage Center in Tucson. Lori has gathered a great amount of information about the Rangers, and she generously made her files available to me. She also took time from her duties on several occasions to seek out and forward answers to miscellaneous questions which arose. It would have been exceedingly difficult to complete this project without her gracious assistance.

The entire staff of the Arizona Heritage Center was helpful to me. I am especially grateful to Joan M. Metzger, researcher for the center's photographic department, who tracked down what must have seemed an endless number of old photos. Unless otherwise indicated, all photos were obtained through this collection, courtesy of Arizona Historical Society.

At the Arizona State Archives in Phoenix, Wilma Smallwood, Carol Downey, and Shirley Macias were most cooperative. John Payne, a member of the U.S. Customs Service who lives in Bisbee, provided fascinating information during a midnight interview in Naco. He served as an intermediary with Fred Valenzuela, who knew Jeff Kidder and other Rangers during his boyhood. John entrusted to a stranger several previously unpublished photos from his private collection.

Scott Pace of Solomon, Arizona, offered information about early-day Solomonville, while Wally Zearing of Benson revealed intriguing tidbits about Burt Alvord and Billy Stiles. Wallace Clayton, publisher of the Tombstone *Epitaph*, provided welcome advice.

Dr. Larry Ball, professor of history at Arkansas State University, is a resourceful scholar of law enforcement in Arizona and New Mexico Territories, and he revealed a number of sources which greatly facilitated my efforts. George Virgines, a historian and writer from Elmhurst, Illinois, provided useful information about Ranger badges and weapons. Also generous with the fruits of their Ranger research were Ronald C. Van Raalte and Fred R. Egloff, from Chicago and Wilmette, Illinois, respectively.

Sgt. Joe Hall of the Marshall (Texas) Police Department offered thoughtful evaluations of Ranger training and arrest techniques. Alan Moore, a longtime officer and instructor of law enforcement, also provided reflective comments. Janice Reece, librarian of the Texas Ranger Hall of Fame in Waco, offered able guidance at her handsome institution.

Melissa Locke Roberts edited the manuscript with remarkable attention to detail, providing valuable suggestions and saving me from numerous errors.

Joyce Chapman, head librarian of Panola Junior College, patiently secured book after book for me through interlibrary loan. Another colleague, J. R. Brannon of the PJC biology department, shared his impressive knowledge of firearms with me; whatever errors I have committed about weaponry occurred because I did not ask J. R. My secretary, Janna Ponder, provided much of the tedious labor of indexing this book.

My wife Faye offered her customary support during this project. While I ventured to Arizona on a long research trip, she presided over our home and four daughters. In my vicarious ride with the Rangers during the past couple of years, she maintained the quality of our family life and otherwise earned my loving gratitude.

# Preface

I regard the Arizona Ranger experience as the last sustained adventure of the Old West. The Ranger company was created in 1901, more than a decade after the census of 1890 had pronounced the frontier closed. But for more than seven years the Rangers galloped across mountains and deserts in pursuit of cattle rustlers and horse thieves, and, blazing away with Colt revolvers and Winchesters, they shot it out with desperadoes in saloons, dusty streets, and desolate badlands.

While most Americans entering the twentieth century were preoccupied with its new developments — riding in horseless carriages, electric trolley cars, and even rickety flying machines; listening to phonographs and viewing motion pictures in nickelodeons; moving to cities and working in factories — Arizonans, inhabiting a vast and hostile wilderness, were concerned about badmen who found other regions of America too settled for their nefarious activities. From late in 1901 until early in 1909, outlawry in Arizona was countered by a hard-riding, quick-triggered little band of men who carried into a new century the traditions of Pat Garrett, Wild Bill Hickok, Commodore Perry Owens, and other gunfighting members of an earlier generation of frontier lawmen. Ranger Captains Burt Mossman, Tom Rynning, and Harry Wheeler, Sergeants Jeff Kidder, J. T. Holmes, and Billy Speed — these and other tough adventurers stubbornly combated badmen in a frontier environment that vanished forever, in great part because of the exploits of the Arizona Rangers.

For the past several years it has been my keen pleasure to follow the trail of the Arizona Rangers, visiting towns where Rangers served and camping in remote areas where the Rangers rode and fought. I saw a few old buildings from the Ranger era in Arizona, admired the mountain scenery and wide open skies they enjoyed, heard the same night sounds. I spent a year reading the daily newspapers of Arizona from 1901 to 1909, discovering numerous stories about the Rangers that have never been published. At the state archives in Phoenix I was permitted to photocopy the remaining records — over 600 pages — of the Rangers, and I also photocopied hundreds of pages of related materials at the Arizona Heritage Center in Tucson.

Curiously, no book has previously explored the rich history of the Arizona Rangers. In 1936 Mulford Winsor, who was acquainted first-hand with the Ranger story, wrote an article about the company for *Our Sheriff and Police Journal*. Joseph Miller's *The Arizona Rangers* (Hastings House, 1972) is primarily a collection of contemporary newspaper stories about the Rangers, and he wrote a fine chapter on the Rangers in *Arizona, The Grand Canyon State* (Hastings House, 1966). There is mention, although often inaccurate, of miscellaneous Ranger incidents in various other books. This is the first attempt to recount the complete history of a hard-nosed, capable, and frequently controversial band of western lawmen. It was fascinating to learn their story.

# 1901:
# A New Force
# in Outlaw
# Territory

*"Give this dollar to my wife. It, and the month's wages coming to me, will be all she'll ever have."* — Pvt. Carlos Tafolla

There were telephones now throughout the West, along with ice cream parlors, bicycles built for two, Coca-Cola, hot dogs, and toothpaste in a tube. Adventurous young men who yearned to follow frontier ways had few places to "hear the owl hoot" at the turn of the century, but Arizona still offered far horizons and a sense of freedom and exuberance. Mountain trails and open rangeland invited those who wished to ride; deer, antelope, bear, and big cats abounded for the hunt. Cowboys still could find large ranches that were hiring, while dishonest drovers stood a good chance of making off with rustled cattle. The Arizona population was sparse: in 1900 the largest town, Tucson, held only 7,531 people, and just 123,000 persons were scattered across the sweeping vastness of the territory. Bank and train robbers, murderers, rustlers, and other lawbreakers with a fast horse stood a reasonable chance of remaining free from arrest.

Individuals with the instincts of a manhunter could find a rare challenge remaining in Arizona. A man could pin on a badge, climb into the saddle and, in the righteous cause of justice and the territorial statutes, gallop into the mountains and canyons and deserts in pursuit of society's enemies. There were still plenty of wrongs to right in Arizona.

On February 15, 1900, for example, a northbound train pulled into Fairbank at dusk and was jumped by five bandits. Express messenger Jeff

1

Milton was hurled off his feet by a severe arm wound, but he scrambled for a shotgun and blasted "Three-Fingered" Jack Dunlap with eleven buckshot. Gunfire continued, and by the time the outlaws forced the door, Milton had fainted from loss of blood. He had hidden the keys before he lost consciousness, however, and the outlaws rode away empty-handed from the scene of the shootout. The heroic Milton was hospitalized for months, and his arm was permanently crippled.[1]

In March, cattle rustlers murdered Gus Gibbons and Frank Lesueur northeast of St. Johns. Veteran law officers George Scarborough and Walter Birchfield trailed the gang into a canyon, but a rifle bullet ripped through Scarborough's leg and killed his horse. Scarborough was taken to Deming, New Mexico, where his leg was amputated, but he died the day after being shot, on April 6, 1900.[2]

Two days later, at noon, Billy Stiles, a former lawman-turned-train robber, confronted jailer George Bravin in Tombstone's Cochise County Courthouse. Stiles, who had been given trusty privileges following a confession, produced a revolver and demanded the keys. Bravin resisted, and after a brief scuffle he was shot in the leg by Stiles, who then tried to release all of the prisoners. Only two men followed him to freedom. A couple of weeks later, the fugitives sent the jail keys back to Tombstone along with an arrogant note: they had met some men who were wanted for killing a gambler, but they did not arrest the "killers" because "we had no warrant and were afraid we could not collect the mileage." Stiles was in and out of custody again, but early in 1901 he was captured by Sheriff Del Lewis while in a Casa Grande saloon. After being handcuffed, however, Stiles bolted for the rear door and escaped into the darkness.[3]

With outlaws seemingly gaining the upper hand in Arizona, cattlemen, mine owners, railroad officials, and newspaper editors pressured Territorial Governor Nathan Oakes Murphy to combat lawlessness with a special force modeled upon the famed Texas Rangers. As early as 1898 an editorial proclaimed the need for a band of rangers: "When such conditions exist, a company of paid 'Rangers' are required to stamp out and destroy the characters that bring about such a state of affairs. Let us have a Territorial Ranger Service."[4]

Governor Murphy was fully aware of the problem. By 1901 he was convinced of the need for a band of rangers. Visiting with rancher-businessman (and fellow Republican) Burt Mossman in Phoenix, the governor stated, "What we need is a hard riding, sure shooting outfit something like the Texas Rangers or the Mexican Rurales."[5]

Mossman had earned a reputation for successfully battling rustlers, and the governor asked him to work with Frank Cox, head attorney for the Southern Pacific Railroad, in outlining a ranger organization. Cox and Mossman repaired to Burt's room in the Adams Hotel, where they worked until midnight hammering out the details of a proposed ranger force. Later they polished their eighteen-point document, then presented it to Governor Murphy.

*Governor Nathan Oakes Murphy, who created the Arizona Rangers through clever political maneuvering.*

(Courtesy Arizona Historical Society, as are photos hereinafter uncredited)

Opposition to spending tax money was a threat to any public project in Arizona, and certain county sheriffs and local officers could be expected to resist a police force with territorywide authority. Governor Murphy, therefore, discreetly worked behind the scenes with rancher-legislators and other members of the Council and House who supported a ranger force. Legislators such as Richard Gibbons, an Apache County cattleman seriously victimized by rustlers and the uncle of murdered Gus Gibbons, could be expected to cooperate readily with the governor.

In March 1901, as the Twenty-first Legislative Assembly met for its final days in session, the ranger bill was maneuvered through committees without fanfare. No newspapers trumpeted the cause, no politicians orated for the rangers, no lobby groups espoused the governor's project. On March 13, 1901, just eight days before the close of the session, Representative Gibbons was given unanimous consent to introduce the bill. Strong support promptly was voiced for the proposed ranger company, while little opposition developed. One representative, newspaperman Kean St. Charles of Mohave County, grumped that the primary activity of all the rangers he had known was "to go camping around the country making coffee," but when the issue came to a vote St. Charles added his support. Despite the usual crowded calendar at end of session, or perhaps because of this, the bill passed on March 21. Governor Murphy wasted no time in adding his signature.[6]

The bill created a fourteen-man force: one captain, one sergeant, and twelve privates. The captain would be paid $120 per month; the sergeant $75 monthly; and the privates $55 each. The men were to provide their own arms, mounts, "and all necessary accoutrements and camp equipage," although the territory would replace horses killed in action. Rangers were authorized to temporarily confiscate horses, if needed, while in pursuit of criminals. The territory would provide ammunition, food, and forage for each Ranger, not to exceed one dollar for meals and fifty cents for horse feed per day. A tax of five cents on every $100 of taxable property in the territory was to be collected and placed in "the ranger fund," which would finance all necessary expenses.[7]

Rangers were to be enrolled for twelve-month enlistments, and they would be exempted from military or jury service. The force was to "be governed by the rules and regulations of the army of the United States, as far as the same shall be applicable." Rangers were empowered to arrest lawbreakers anywhere in the territory, then deliver prisoners to the nearest peace officer in the county where the crime was committed.

The governor was authorized to appoint "competent persons as captain and sergeant," and Murphy knew exactly whom he wanted to head the Arizona Rangers. Burton C. Mossman had impressed most people since his arrival in Arizona in 1893.

Born in 1867, Burt Mossman was the son of an Illinois farmer who rose from private to major with the Thirty-sixth Illinois Volunteers during the Civil War. When Burt was a boy the family moved to Missouri, then migrated to New Mexico in 1882. He learned to speak Spanish, worked with a survey crew in the Apache-infested Sacramento Mountains, and hiked 110 miles round-trip across the desert to a stage stop with surveyors' reports for Washington.[8]

At seventeen Mossman began cowboying, but a fiery temper cost him four jobs. Nevertheless, while still a teenaged cowboy he won appointment as a roundup rep, and he became a foreman at the age of twenty-one. Five years later he was employed to manage Arizona's Bar OO spread along the Verde River north of Phoenix.

Mossman found it convenient and pleasant to spend his off-hours in the capital city. The stocky young cattleman was congenial and a good storyteller. He was fond of poker, cigars, good liquor, and the company of other vigorous, imposing men. Despite his winning personality, his temper remained short-fused and he was a dangerous man to cross. In 1896, while on a trip to Mexico, he quarreled with a young Mexican captain in a Mazatlán *cantina* and was challenged to a duel at dawn. The principals loaded their pistols each with a single bullet. They stepped off fifteen paces, turned, and the Mexican fired his German Luger. The bullet missed, and Mossman pumped the .45 slug from his short-barreled Colt into the captain's shoulder. At that point a squad of local police arrived,

*The Adams Hotel in Phoenix, where Burt Mossman and Frank Cox worked to outline
a Ranger organization.*

No. _____    $ _____

### Ꚍꜱ the Territorial Treasurer of Arizona Territory

*Phoenix, Arizona,*

*Pay to the order of* _____

_____ *Dollars,*

*Account of* _____

COUNTERSIGNED:

*A Ranger fund warrant, through which expenses of the company were paid.*

(Author's collection)

and Mossman spent four weeks in the *calabozo* until he managed to escape with the help of a Mexican friend.

In 1897 the Bar OO was sold, but by year's end Mossman was employed as superintendent of the vast Aztec Land and Cattle Company. The Hash Knife, as the famous spread was known, grazed 50,000 cattle and 2,000 horses on a two-million-acre range located eighty miles west of Holbrook and forty miles south of the Atlantic and Pacific Railroad tracks. Under negligent management, the Hash Knife had become a haven for lazy and often dishonest cowboys, many of whom actually helped rustlers like the notorious Bill Smith and his gang. For fourteen years in a row, the Hash Knife had been unable to pay a dividend to its investors.

Mossman realized he had to halt rustling. When he arrived by rail at Holbrook in January 1898, he determined to take prompt action. Disembarking from the train in a suit and derby hat, he was met by cowboy Charlie Fought, who was to take the new manager to headquarters eleven miles away. Instead, Mossman had Fought guide him to a stolen herd penned thirty miles from Holbrook. Mossman and Fought rode into camp like wanderers seeking a fire in the winter cold, but Burt dismounted and jammed a six-gun into the gut of one thief. The three men surrendered. When they were delivered to Holbrook the newly elected sheriff, Frank Wattron, was sufficiently impressed to swear-in Mossman as a deputy.

Mossman rode the Hash Knife ranges for a week, familiarizing himself with the land, livestock, and men. Then he fired his foreman and fifty-two of his eighty-four riders. During the next several months, he sent a dozen rustlers to Yuma Territorial Prison, shipped ten carloads of cattle for sale, and branded 16,000 calves. He declined Bucky O'Neill's offer of a Rough Rider commission so that he could concentrate on his new position.

But in 1900 the Aztec Land and Cattle Company, like other western ranching syndicates, decided to liquidate their holdings. Mossman and a partner bought out a Bisbee slaughterhouse and retail shop. Burt opened a branch in the new border town of Douglas, erecting a concrete building that he soon would sell for a handsome profit. Col. William C. "Bill" Greene, a dynamic promoter, copper magnate, financier, and rancher, tried to work out a partnership arrangement with Mossman to manage a ranch north of Willcox, but the deal did not materialize.

In 1901 Mossman moved to Phoenix, his profits from the slaughterhouse enabling him to be in no hurry to find a new occupation. He took a room at the Adams Hotel, leisurely indulging in poker and the company of his friends, several of whom tried to persuade him to head up the new ranger force. Mossman later stated that he twice declined the position.

Finally, Governor Murphy summoned Mossman to his second-floor office in the splendid new capitol, which had been formally dedicated in February 1901. When Mossman entered the office he was confronted by a lineup of influential friends and poker-playing cronies: Governor Mur-

*Burt Mossman in later years. At thirty-four he became the first captain of the Arizona Rangers.*

phy; Colonel Greene; Col. Epes Randolph, head of Southern Pacific Railroad operations in the Southwest; Charley Shannon, Council member from Graham County; and J. C. Adams, once an important Chicago attorney and now the proprietor of the Adams Hotel. Within an hour this array of powerful, persuasive men had talked Mossman into accepting the captaincy of the Arizona Rangers. Mossman stipulated that he would serve just one year, and he extracted concessions that he could select his own successor and "that I would not be interfered with." [9]

Mossman's commission was dated August 30, 1901. Section 10 of the Ranger Act stated that the captain should select "as his base the most unprotected and exposed settlement of the frontier." Captain Mossman decided that Ranger headquarters should be in Bisbee, the recent site of his meat-packing business.

Located eight miles north of the Mexican border in southeastern Arizona, Bisbee was a raw mining town in the midst of the territory's most lawless area. Bisbee was the principal community in the wealthy Warren Mining District, centered on the six-mile-long canyon called Mule Pass Gulch. Copper traces had been spotted in Mule Pass Gulch as early as 1875, and by the 1880s the Copper Queen Consolidated Mining Company had begun to exploit ore deposits which in time would produce more than $100 million in profits.

Hardcases from Tombstone gravitated to Bisbee during the 1880s, giving the mining camp a tough reputation. Brick buildings lined the

*The Arizona State Capitol, where Burt Mossman was persuaded to assume command of the Rangers.*

main street, which wound for two miles along the steep upper end of the canyon. Ramshackle frame structures teetered above the street, perched uncertainly on rocky slopes. Probing outward from the main thoroughfare were side streets that curled into rocky ravines. By 1901 mining officials and engineers from the East had erected Victorian residences throughout the canyon, while saloons, dance halls, gambling houses, and brothels lined mile-long Brewery Gulch. Mossman rented office space for the first Ranger headquarters in a building that also housed a Bisbee justice of the peace.[10]

Mossman now had to recruit thirteen Rangers. He wanted outdoorsmen — men who could ride and trail and shoot, men who had experience as cowboys or peace officers. He had several candidates in mind, but he went about the enlistment process quietly. Mossman did not announce the new members of the force, hoping to guard their identities from lawbreakers.

Indeed, throughout the history of the Rangers, it remained customary not to broadcast the identity of company members. Several Rangers became famous throughout the territory, and some men were stationed for long periods of time in the same community and thus were well known locally. But there was to be considerable turnover among Ranger personnel, and many new men concealed their identities, while several men performed undercover assignments. Certainly, it was intended that the charter Rangers would not be widely known.

*Bisbee, the raucous copper mining town which became the first headquarters of the Arizona Rangers.* (Author's collection)

The first men enlisted were Bert Grover and Tom Holland, who signed up at headquarters in Bisbee on Friday, September 6.[11] Joining the force that same day were Texans Leonard Page and George Edgar Scarborough, who signed his papers in Tucson. Ed Scarborough, a mere twenty-three years old, was the son of George W. Scarborough, the veteran officer who had died the previous year after taking a rifle bullet from a rustler near San Simon.

The next week Carlos Tafolla enlisted at St. Johns. On September 13 Fred Barefoot and James Warren signed up at Clifton, along with Texan Don Johnson, at twenty-two the youngest of the original Rangers. Five days later John Campbell, a native of Pittsburgh, Pennsylvania, and Richard Stanton, from New York City, were accepted in Phoenix. Campbell, although most recently a farmer, had amassed "unusual military experience" and eventually rose to sergeant. Enlisting on September 20 was Duane Hamblin, a tanned, rugged-looking outdoorsman with a sweeping mustache and a growing family.

The thirteen vacancies were filled in October, with Henry Gray, a forty-seven-year-old Californian and the oldest of the charter Rangers, as the twelfth, and forty-year-old Frank Richardson as the thirteenth.

The average age of the Rangers was thirty-three. Three men were in their twenties, three in their forties, and the other seven, along with Cap-

*A family portrait taken at the Page Ranch near Willcox in 1910. On the top row, sec-
ond from right, stands Leonard Page, one of the first men enlisted as an Arizona
Ranger.*

tain Mossman, were in their thirties. Scarborough, Johnson, Tafolla, and
Grover had previous experience as peace officers, while seven of the men
listed their prior occupations as cowboys or cattlemen. New Yorker Richard Stanton stated that he was a waiter by profession, but he had fought
with the Rough Riders in Cuba and Mossman thought he saw the qualities of a Ranger in him. Within a few weeks the captain realized that he
had erred in hiring Stanton.

Several Rangers were in camp at Naco on Friday night, November
15. Dick Stanton wrangled with Tom Holland, who angrily turned away
as Bert Grover tried to intervene. Grover demanded that the trouble stop
and threatened to place Stanton under arrest. Holland stomped off,
whereupon Stanton turned his anger on Grover. Stanton furiously
clutched his revolver, but Leonard Page stepped in to stop his draw. Now
aroused, Grover pulled his six-gun and leveled it at Stanton. Ignoring
Grover's drawn revolver, Stanton went for his Winchester and tried to
lever a shell into the chamber. But Grover was on top of him, wrestling
him to the ground and again leveling his six-gun. In the scuffle, however,
someone struck Grover's hand downward and his revolver discharged.
The bullet took a plug out of Grover's right leg.[12]

At the sound of the shot the Rangers came to their senses. Grover's
wound proved superficial, but he was confined to his bed for a couple of
days. Although the men patched up their touchy relationships, Mossman
intended to have no further dissension. On December 3 the contentious

*The Arizona Militia of Duncan, an 1885 Indian-fighting unit. The stocky man standing at far right is a younger Frank Richardson, who, at the age of forty, became a charter member of the Arizona Rangers.*

Stanton was terminated, his "services no longer required" by the Rangers.

Democrats complained to Governor Murphy that the Ranger appointees were all Republicans, and Captain Mossman was called to appear in Phoenix. The governor relayed the complaints and suggested that Mossman find some Democrats.

Mossman, although a staunch Republican, had never asked his men about their politics, being far more concerned with their ability to ride and shoot. "Now governor," said Mossman, "if you think I can go out in these mountains and catch train robbers and cattle rustlers with a bunch of Sunday School teachers, you are very much mistaken.

"As it is now," continued Mossman, firmly imbedding his tongue in his cheek, "every time one of my men gets killed, he's a Democrat, and there are too many of them in this territory already. So, all I have to do is keep appointing Democrats and there soon won't be any left to worry the Republicans." [13]

Mossman reminded his superior that he had been promised a free hand in organizing the Rangers. Governor Murphy finally laughed off the whole matter, and Mossman remained unfettered in running his force. But the political issue did not vanish, and Arizona Democrats continued to regard the company as a Republican assemblage — a point of view which would haunt the future of the Rangers.

Section 4 of the Ranger Act had specified that the men were to use "the most effective and breech-loading cavalry arms." The territory would purchase the weapons, but each Ranger was to pay the cost of his rifle "out of the first money due him." Captain Mossman knew the rifle he wanted his men to carry. When he was chasing rustlers on the Hash Knife in 1898, Mossman had ordered three Model 1895 .30-.40 Winchesters from St. Louis. After these splendid weapons arrived, Burt presented one to Sheriff Frank Wattron and another to Deputy Sheriff Joe Bargeman. Mossman tried out his powerful gun by sending a bullet through a cottonwood tree that measured 103 inches in circumference.

John M. Browning, America's foremost genius in arms design, devised the Model 1895, which was a landmark in Winchester production. The 1895 Winchester was the first lever-action repeater to use a box magazine instead of the old tubular magazine. Five rounds nestled in the box, and the chamber could accommodate a sixth. One advantage of this rifle was that it used the same caliber cartridge as the Army Krag, and "we could always be sure the commanding officers at Fort Huachuca and Fort Apache would load us up with plenty of ammunition whenever we ran low." [14]

Each member of the company was expected to furnish his own "six shooting pistol (army size)." The Rangers thus were armed with powerful Winchester repeating rifles and the reliable Colt .45 single-action revolvers. The territory would furnish ammunition, along with provisions and forage for horses.

There was to be no uniform. Like the Texas Rangers, members of the company would outfit themselves in range clothes, not only for durability but also to insure the anonymity desired by Mossman. Under Mossman the Rangers would not even wear a badge.

During the first month of their existence, Ranger expenses totaled less than $600. Salaries for the incomplete force were $434.65, while provisions, ammunition, and forage added up to just $137.70. In addition, Mossman spent $18.85 on "sundries." From a Prescott firm he bought four pack mules: three for $60 each and one for $40. Total expenditures for September 1901 were $591.20 — an amount which would not even cover salaries when the roster was full.

The next month, when enlistments were completed and operations against outlaws had begun, provisions, ammunition, and forage cost the Ranger fund $320.20. In November this figure increased to $459.97, and in the last month of 1901 the territory spent $556.38 to keep the Rangers in the field. In addition, Mossman in December spent $254 on pack animals, packsaddles, and other needed "furnishings." [15]

The first big Ranger fight erupted before Mossman could complete his thirteen-man roster.[16] An immediate target of the Rangers was the Bill Smith Gang. This band of rustlers made their headquarters in northern Graham County, where Bill Smith and his younger brothers and sister lived at their mother's home on the Blue River, near Harper's Mill.

As a young man, Bill had drifted into Oklahoma Territory, where reportedly he served an apprenticeship in rustling and other frontier chicanery with the Dalton brothers. By the turn of the century, he was Arizona's most notorious cattle rustler. Bill had first been arrested in 1898 for cutting out and weaning a score of calves from ranchers Henry Barrett and Bill Phelps. Smith was jailed at St. Johns. Jailer Tom Berry found Smith to be such a hard sleeper that it was necessary to enter his cell at breakfast time to awaken him. But Berry was merely the victim of an escape ruse. One morning when he walked into the cell, Berry found himself staring into the muzzle of a .45 revolver, smuggled to Smith by his brother Al. Smith locked Berry inside the cell, then slipped out to a woodshed, where Al had left a Winchester and ammunition.

After his escape, Smith fled to New Mexico for a year. When he returned to Arizona he was wanted for train robbery, and he continued to steal cattle and horses, often from Henry Barrett. During the first week of October 1901, the Bill Smith Gang was spotted heading south near Springerville with a herd of stolen horses. A couple of days later, one of the younger Smith brothers rode into St. Johns to buy supplies, and he casually asked the whereabouts of Barrett. The tough old rancher heard of the inquiry and proceeded to organize a posse. Barrett rode with Hank Sharp, Pete Peterson, and Elijah Holgate to Greer, where they found Rangers Carlos Tafolla and Duane Hamblin. Tafolla and Hamblin, having been assigned to search for the Smith Gang, readily joined Barrett.

The posse followed the trail three miles south to Sheep Crossing of the Little Colorado River and on to Lorenzo Crosby's ranch on the Black River. There the posse enlisted Crosby and the Maxwell brothers, Bill and Arch. The Maxwells, regarded as superb scouts, had been friends with the Smiths until the gang made off with several horses from the Maxwell range.

The rustlers' trail led south to Big Lake to Dead Man's Crossing on the Black River to Pete Slaughter's ranch where, according to the signs, the gang had camped. The posse pitched camp at the same site, then the next day, October 8, followed the trail six miles down the west bank of the Black River.

There is no more beautiful or forbidding wilderness in America than the Black River country. In October the temperatures are crisp during the day and frigid at night, and the forests are a riot of orange and red leaves, with a thick carpet of pine needles on the ground. Soaring mountains bristle with towering pine and spruce trees, cedars and junipers. It is difficult country to traverse: the narrow, winding valleys are too thickly forested for easy passage, while the steep mountainsides offer treacherous, boulder-strewn angles littered with fallen timber. Breaks in the timber from high on the mountains reveal breathtaking views of wild beauty, but the almost impenetrable wilderness provided a natural hideout for fugitives. Rustlers regularly found refuge in the area, and the nearby western border of New Mexico offered an additional avenue of escape.

On Tuesday, October 8, the outlaws were in camp at Reservation

Creek, in a gorge 200 yards wide and 100 feet deep near the headwaters of Black River. The gang had shot a bear and late that cold afternoon were engaged in skinning the beast. Some of the gang members had started supper, while bloodhounds prowled the camp perimeter. One hound nervously barked out an alarm, and Bill Smith scrambled to the top of a rim for a look. He darted back to camp with news that several men were approaching. Al and George Smith began to move the horses out of the clearing.

The posse had heard the trio of rifle shots that brought down the bear, and a ride of half a mile brought them to a bloody trail in the snow. Sign indicated that two men were packing out freshly killed game on a horse, and the posse, sensing their prey, followed the trail in the final hour of daylight.

The nine posse members tied their horses to a cluster of bushes and crept the last 300 yards through the snow on foot. They moved in from the west as the sun set between Mount Ord and Old Baldy. The outlaws thus enjoyed the protection of a shadowed gorge, while sun rays, highlighting the rim to the east, made it difficult to fire into the rustlers' camp. Most of the possemen crawled to prone positions on the rim, but the two Rangers and Bill Maxwell boldly advanced into the clearing. In the open, they were starkly silhouetted against the whiteness of the snow.

Barrett shouted from the rim for the lead man to get down. Hamblin flattened onto the ground, but Tafolla and Maxwell ignored their danger. Maxwell called out an order to surrender.

"All right," replied Bill Smith. "Which way do you want us to come out?"

"Come right out this way," directed Maxwell.[17]

The outlaw leader walked toward the lawmen, dragging a new Savage .303 rifle behind him. Suddenly, Smith brought up the Savage repeater and opened fire from a distance of forty feet. Tafolla went down, shot through the torso, while Maxwell was hit in the forehead and died on the spot. Smith darted for cover as the other outlaws began firing from behind tree trunks. Tafolla gamely emptied his Winchester, and his companions opened up from the rim. Most of the rifles were loaded with black powder cartridges, and a haze of white smoke began to spread through the gorge as gunshots echoed off the surrounding walls. Barrett's fire was especially effective; the rancher was armed with a Spanish Mauser captured in Cuba, and the smokeless, steel-jacketed rounds ripped through the little pine trees that shielded the outlaws. Two rustlers were wounded, shot in the foot and leg, and one of their hounds was killed. After a few moments, the gunfire ended as the gang retreated into the timber.

During the shooting, Hamblin had worked his way around to the outlaw mounts. He found nine saddle horses and a pack mule, drove the animals away, and put the rustlers afoot. Desperately, Smith and his men pressed into the wilderness and escaped into the sudden mountain nightfall.

*Duane Hamblin with his family. Unlike his Ranger companion, Carlos Tafolla, Hamblin survived the fight with the Bill Smith Gang.*

Back in the clearing Tafolla lay on his back, shot twice through the middle and moaning for water. Bill Maxwell was dead; his big hat had three bullet holes in the crown. As the posse closed in they found the dead hound, along with saddles, bridles, camp gear, and personal belongings that the fleeing outlaws had abandoned. Tree trunks throughout the gorge were scarred with bullet marks. The clearing forty miles south of St. Johns would become known as the Battle Ground.

Bill Maxwell's hat was left on the ground, and cowboys who later had occasion to ride through the Battle Ground superstitiously refused to touch the bullet-riddled *sombrero*. Maxwell was carried to where the posse had tethered their horses and was laid out on two saddle blankets. Hank Sharp and Arch Maxwell rode east for help to the Mormon community of Nutrioso, where the Maxwell homestead was located. Hamblin, Barrett, Peterson, Holgate, and Crosby stayed behind to provide crude care for the agonized Tafolla. Before he lost consciousness Carlos, realizing that he was dying, pulled a silver dollar from his pants pocket and handed it to Barrett.

"Give this dollar to my wife," gasped Tafolla. "It, and the month's wages coming to me, will be all she'll ever have." [18]

Tafolla died at midnight. Captain Mossman received word at Solomonville of the tragic fight. The message had been sent by Henry Hunig, a St. Johns merchant and sheepman. While he was readying his horses, Mossman received a second telegram: "Tefio [*sic*] died from wound send force if possible up Blue to Mrs. Smiths place near Harpers Mill site she is mother of one of the murderers." [19]

Mossman sent orders south to guard the routes into Mexico, then he rode out with three Rangers. They spent the first night at Clifton and reached the Battle Ground late the next day, where Henry Barrett described the fight. From the San Carlos Indian Agency the captain acquired two Apache trackers, Josh and Chicken, and Mossman and his men plunged into the White Mountains in search of the Smith Gang.

Bill Smith had led his men through the snow-covered wilderness. On Beaver Creek the gang roused a slumbering camp of drovers and obtained a meal. The cowboys had heard of the fight and revealed the identity of the dead men.

"Well, I'm sure, sure sorry," said the outlaw leader after learning that Bill Maxwell had been slain. "When he stood up that way we thought it was Barrett, he was the man we wanted. We feel mighty sorry over killing Bill Maxwell, he was a good friend of ours. Tell Bill's mother for us that we're very sorry we killed him." [20]

The fugitives continued on into Bear Valley between the Blue River and the New Mexico border. They arrived at the isolated ranch of Hugh McKean and asked to buy horses. When McKean refused, they rounded up his best mounts and saddles, seized his guns and a sack of food, then headed toward New Mexico.

Mossman's Apache trackers led the Rangers to the McKean ranch the next day. Mossman pressed on, but a heavy snow descended and obliterated the trail. Another snowstorm drove Mossman back to the Mc-Kean ranch, but stubbornly he rode into New Mexico. Josh and Chicken picked up a trail which led to Magdalena and to the Rio Grande three miles south of Socorro. At this point the trail disappeared and Mossman returned to Arizona.

Four other Rangers had made their own sweep into New Mexico. Bert Grover, Leonard Page, Dick Stanton, and Tom Holland were in Naco when orders from Mossman arrived by telegram (Grover became miffed when "at least fifteen people told him the contents of the telegram before he received it").[21] The four Rangers rode out of Naco on the night of October 11 and headed east. Finding no sign of the Smith Gang along the Mexican border, the Rangers moved into New Mexico through heavy rainstorms. They found no one to arrest, but they did round up a number of horses that had been stolen from various Arizona ranches. The four returned on Thursday, October 24, saddle weary and with no prisoners.

A favorite haunt of rustlers had been penetrated, so word began to spread among outlaws. The Smith Gang never resumed its activities in Arizona. The widow Smith later told Ranger Joe Pearce that Bill and Al made their way to Galveston, Texas, where they took a boat to Argentina. In 1909 George Smith returned and surrendered himself to Sheriff Jim

Parks of Graham County. However, since the only charges against him had been filed in Apache County, he was released from custody. George Smith settled at his mother's ranch on the Blue River and tended to her little herd of cattle.

When Henry Barrett told Captain Mossman about Carlos Tafolla's silver dollar and dying remarks, the captain asked for the coin. On his next visit to Phoenix he reported personally to Governor Murphy, describing the fight at the Battle Ground and the fruitless pursuit into New Mexico. Mossman then pointed out that Tafolla had left a penniless widow with three children at St. Johns. Tafolla's paycheck for October amounted to just $14.66 because he had died on the eighth day of the month; his September paycheck, for his first twenty-one days as a Ranger, had been for only $38.49. For services to Arizona which cost him his life, Carlos Tafolla had been paid a total of $53.15. Mossman asked the governor for a pension for Aceana Tafolla, and that day Murphy called the appropriate committee together for a special session. Mossman presented his plea, and the legislature approved a pension of $25 a month for two years.[22] The next two legislatures, however, cut the pension in half to $12.50 a month. The final territorial legislature was a bit less heartless, appropriating $20 a month for the family of the dead Ranger.

The captain visited Mrs. Tafolla in St. Johns, telling her about the pension and presenting her with the silver dollar. Tafolla had been buried with such ceremonies as could be mustered.

The Bill Smith Gang was not the only one to shoot their way past the Rangers; few arrests were made by the new law enforcement agency in 1901. The first Ranger apprehension was recorded on October 2, when suspected murderers Andrew Griffin and Hete O'Connor were captured in the Huachuca Mountains. A week later, horse thief Frank Hollis was arrested as the Rangers scoured the Black River country. But the Rangers, preoccupied with a fruitless search for the killers of Carlos Tafolla and Bill Maxwell, recorded no further arrests for more than a month.[23]

On November 11 cattle rustlers James Head and William Williams were apprehended in the Chiricahua Mountains. For several months ranchers in the Chiricahua and Swissholm Mountains had regularly lost cattle. Some of the animals had been stolen, while others had been killed on the range. The brand would be cut from the hide of a slain animal, a side of beef would be butchered, and the rest of the carcass left to rot. Responding to complaints, Captain Mossman detailed Bert Grover and Leonard Page to investigate. Grover and Page prowled around the area for several days with no result, but on Monday, November 11, they rode into Hunt Canyon in the Chiricahuas. Ahead of them two men later identified as Head and Williams roped and killed a steer. The two Rangers concealed themselves as Head and Williams cut off the brand and began to butcher a hindquarter. The Rangers then moved in, arrested Head and Williams at gunpoint, and shackled the two rustlers. After an overnight stay at a ranch, the Rangers took their prisoners in to Bisbee, where Head and Williams were quickly tried and convicted.[24]

Head and Williams, however, were the only lawbreakers captured during November. In December the Rangers were briefly back at full force: Texas cowboy McDonald Robinson filled Tafolla's vacancy on the first of the month, although two days later Richard Stanton was discharged. But there seemed to be little improvement in arrest totals. On December 5 Martin Woods was arrested at San Carlos, and five days later John Ruth was taken into custody at Cochise; both men were charged with grand larceny.

During the first three months of their existence, the Rangers had managed to apprehend merely seven malefactors, and the only major clash with outlaws ended tragically for the new organization. On the last day of 1901, however, the Rangers moved in on four murderers. Cruz Figuerra, Ramón Moreno, Trinidad Ariola, and Francisco Hernandez were arrested in the Huachuca Mountains on December 31. The Rangers had scored an impressive coup at the close of the year, and it was a sign of things to come for 1902.

# 1902: Tracking Chacón and On to Douglas

*"They ought to thank me for giving them a chance to come in and take their medicine."* — Capt. Burt Mossman

The first Ranger event of 1902 was the appointment, on New Year's Day, of Dayton Graham as sergeant. Because of the death of Carlos Tafolla, the Rangers under Mossman's command had not totaled more than twelve men.[1] For the post of sergeant Mossman had wanted Graham, a Bisbee peace officer, and he held the job open until Graham consented to join the force.

On his fifth day as a Ranger, Graham led a posse that nabbed a trio of rustlers. But after serving just one month, Sergeant Graham was offered the position of city marshal of Bisbee, and he resigned from the Rangers on February 1. Mossman refused to appoint anyone else sergeant. Headquartering in Bisbee, he saw Graham regularly, and after a two-month siege persuaded the former sergeant to reenlist. Graham assumed his old rank on April 1, but Ranger service still held little attraction for him. On May 31 he resigned again, this time for good, and he became a constable in Douglas, where he would find himself involved in a Ranger shooting in 1903.[2]

Two more Rangers resigned in January, and another in February, but while Mossman juggled his roster and searched for tough new recruits, he found himself increasingly busy with arrest reports. During the first week of January, several rustlers were rounded up and turned over to various local officers. More arrests followed, and the apprehension of

*Bisbee's first city government (1902) included Dayton Graham (bottom center). Graham resigned his city marshal's position to accept the sergeancy of the Rangers.*

murderers, rapists, cattle rustlers, and horse thieves early in 1902 proved that the Rangers were warming to their task.[3] But the most impressive Ranger move to date came in March against a contingent of the notorious Musgrave Gang.

George Musgrave and his brother Canley led a band of outlaws wanted in Texas and New Mexico for train robbery and murder. Congregating in New Mexico, they followed the example of many other badmen and regularly slipped across the Arizona-New Mexico line, finding safety from the law in the mountain wilderness of eastern Arizona. Musgrave had even attempted an abortive, but bloody, bank robbery at Nogales.

On January 27, 1902, one of Musgrave's men — probably Joe Roberts — held up the post office at Fort Sumner, New Mexico. The bandit made off with valuables from several citizens, as well as the post office funds. Roberts headed for Arizona, in company with other gang members, who were suspected of robbing a store and killing the proprietor. New Mexico officials notified the Rangers, and Mossman ordered men he had stationed in Clifton to be on the alert for the fugitives.

One of the desperadoes was known to be Witt Neill, and the Rangers learned that he had a sweetheart living in a house near the Blue River about forty miles north of Clifton. Then John Parks, a Graham County deputy sheriff stationed in Clifton, received word that several heavily armed men had been spotted in the vicinity of the woman's house. Parks formed a posse, which included Rangers Fred Barefoot, Henry Gray, and Pollard Pearson. Also riding were special officers Clyde Barber, Dick Boyle, Ed McBride, W. B. Trailer, and Willie Willis, as well as citizens Gus Hobbs, H. D. Keppler, and Frank Richardson, who had resigned from the Rangers on February 28 and lived in Solomonville.

The posse left Clifton on March 9 and rode directly to the cabin of Witt Neill's sweetheart. Arriving late at night, the lawmen surrounded the house and closed in at dawn. They jumped Neill as he slumbered on a bed on the porch. Covered by several officers, the groggy fugitive did not make a move. When his blankets were turned back, an arsenal was revealed: two six-guns, a .30-.40 Winchester, a double-looped belt full of cartridges, and forty or fifty more cartridges stuffed into his pockets.

Three of the officers returned to Clifton with their prisoner, while the rest of the posse backtracked to a trail they had spotted the previous day. Sign showed that two men had ridden west, while three riders went south. A couple of Rangers went after the men who had gone west, and the other seven posse members headed south.

Later in the day the seven-man group caught up with outlaws George Cook and Joe Roberts in the mountains. Cook and Roberts were driving two dozen stolen horses, and as the rustlers concentrated on their work, the posse members split into two parties and surrounded them. Looking down several Winchester barrels, the outlaws prudently submitted to arrest. A man who had been with them had slipped away; a few days later it was reported that he was captured by lawmen in New Mexico east of Carlisle.[4]

Neill, Cook, and Roberts were jailed in Solomonville, the seat of Graham County, and soon they were released into the custody of federal officers for return to New Mexico. Mossman arrived in Solomonville following the arrests and took possession of the outlaws' rifles, pistols, and knives. Later, United States Marshal Creighton M. Foraker of New Mexico requested the weapons for use as evidence, but Mossman declared them "spoils of war" and held onto the outlaw hardware. As late as 1935 he remarked to Pollard Pearson that he still had some of the confiscated weapons.[5]

Pearson, who had joined the Rangers in February, was among those who rode south to capture Cook and Roberts. Years later he offered an intriguing speculation about the two outlaws who had slipped away to the west after being placed in federal custody. In 1936 the skeletons of a man and horse, a saddle marked XIT, and a shotgun were discovered in the Hedgepet Mountains twelve miles north of Glendale, Arizona. Pearson felt strongly that the two outlaws had wandered into treacherous country, and that at least one of the fugitives had perished, perhaps from lack of food or water.[6]

At the close of each fiscal year in Arizona Territory, June 30, a "Report of the Governor of Arizona" was submitted to federal authorities. For the duration of the Rangers' existence, this report included a summation of their activities throughout the preceding twelve months. In the four-paragraph report on the Rangers for 1902, the strongest boast regarded George Musgrave's henchmen: "One of the most noted bands of cattle thieves was broken up . . ., and numbers of small bands of outlaws have been taken into custody or driven from the country. The operations of the Arizona Rangers have been most successful." [7]

Indeed, the Rangers had been busy during the first half of 1902. Four separate arrests during a two-week period of January put seven rustlers behind bars (two horse thieves were seized by Rangers who had slipped across the border into New Mexico's Animas Valley). In February, Cook, Neill, and Roberts had been arrested, along with a murderer at Willcox, two rapists snatched out of La Cananea, Mexico, and assorted horse thieves and burglars.

March brought a mounting number of arrests, and April proved to be the busiest month of the year. Eighteen miscreants were nabbed during April, including J. W. Smith, a thief and murderer who used the nostalgic alias "Sam Bass." Horse thieves and smugglers headed the list of felons seized by the Rangers in April, but on Monday, the seventh day of the month, members of the company participated in a fatal fight with rustlers.

Sheriff Jim Parks of Graham County led a posse, which included Rangers, in search of a band of thieves who were killing and butchering cattle. The officers located their prey at Eagle Creek near Morenci. The rustlers were making jerky from six freshly killed beeves when the posse moved in. There was "some exciting shooting" as the gang tried to fight their way past the lawmen. Rustler Manuel Mendosa was shot to death, whereupon his fellow outlaws submitted to arrest.[8]

May, June, and July found the Rangers continuing to concentrate on rustlers: seventeen men were found and turned over to local authorities for "Cattle Stealing," "Horse Stealing," or "Violating Live Stock Law." These rustlers were arrested in Pinal County, Santa Cruz County, Dos Cabezas, Mammoth, Blue River, Cochise, Safford, Naco, Bisbee, the Huachuca Mountains, the Chiricahua Mountains, and La Cananea, Mexico. Usually the apprehended rustlers were driving stolen cattle or horses, and these animals were returned to their owners through county or local officers. One newspaper was moved to comment that the Rangers "have done much for the cattlemen in different parts of the territory, in apprehending cattle thieves who had followed the occupation of stealing cattle so long that they had come to look upon it as a legitimate business."[9] The rustling fraternity began to realize that a crackdown was in progress.

But much of the goodwill generated by the increased arrest records was temporarily negated by an unsavory incident involving Rangers at Bisbee. On Saturday, August 16, Captain Mossman — always eager for the competition and camaraderie of poker — found a game at the Orient Saloon in Bisbee's Brewery Gulch. Bert Grover sat in, along with three or four other players and a professional gambler who ran a table at the Fish Pond Saloon.

The stakes were considerable, one pot reaching $400. The professional laid out his cards and began to rake in the $400 pile, but Grover growled that the hand had been won dishonestly. When the Fish Pond gambler snapped back a retort (reputedly he called the Ranger a "piss-poor card player and a cocky son-of-a-bitch"),[10] Grover whipped out his revolver. Mossman tried to calm Grover, but Bert flourished the six-gun and grew louder. Ranger Leonard Page came to his captain's rescue, and Mossman and Page hustled the livid Grover away from the table.

At this point two Bisbee policemen named Harrington and Jennings burst into the Orient. Tempers flared, and all five lawmen tumbled into the street in front of the saloon. As in most such encounters there was more wrestling and shoving than fisticuffs, but members of the Saturday night crowd eagerly looked on as the "peace officers" scuffled among themselves. The "Johnson Day picnic" (Brewery Gulch slang for a saloon brawl) finally ended when Harrington and Jennings managed to subdue Grover, and Mossman pulled off Page. Mossman protested but did nothing else as the two policemen marched Grover off to jail, accompanied by a mob of boisterous onlookers. A few hours after Grover's incarceration, Page slipped into the jail, found the cell keys, and assisted his fellow Ranger in vanishing from Bisbee.

Public reaction to the incident was hostile. Reported one newspaper: "The whole affair was, to say the least, disgraceful." A few days later a pe-

*The street angling to the left of the Orpheum Theatre was Bisbee's notorious Brewery Gulch, where the Orient Saloon was located. Rangers playing poker at the Orient became involved in a disreputable brawl.*

tition was circulated denouncing Mossman and calling for his removal, and 200 denizens of Brewery Gulch — tinhorn gamblers, petty thieves, barflies, pimps, prostitutes — eagerly added their signatures. Respectable citizens circulated a counter petition in support of Mossman.[11]

The Ranger captain, however, soon vaulted back into public popularity. The most remarkable feat associated with the Rangers during 1902 involved Mossman's daring, though illegal, scheme to capture Augustín Chacón. "One of the blackest hearted villains that ever operated in the southwest," [12] Chacón was a ruthless *bandido* and killer who was widely blamed for the death of more than two dozen men. On Christmas Eve night, 1895, Chacón and his gang robbed a store near the copper mining town of Morenci. Paul Becker, the storekeeper, was brutally murdered by Chacón with a hunting knife. Chacón's men loaded six packhorses with stolen merchandise and rode into the wilderness.[13]

Sheriff Billy Birchfield of Graham County led a posse, which cornered Chacón and his gang in a box canyon. Pablo Salcido, a posse member who had known Chacón, persuaded Birchfield to let him parley under a flag of truce. Salcido tied a white handkerchief to his barrel, walked to within fifty feet of Chacón, and treacherously was shot to death by the bandit leader. In the fight which followed, Chacón was severely wounded and two of his men were slain. Although Chacón recovered from his wound, he was tried and sentenced to hang in Solomonville for the murder of Becker. A hacksaw was smuggled to him in a Bible, however, and nine nights before he was scheduled to hang, Chacón escaped from the Solomonville jail. He murdered two prospectors in a camp at Eagle Creek near Morenci, looted their provisions, then headed for Mexico.[14]

Using a mountain base in northern Sonora, Chacón and his men crossed frequently into Arizona to rob and kill. Now sporting a black beard, he was called *Peludo* ("Hairy") and his villainous exploits became legendary. *Peludo* was blamed for almost any crime of unknown origin, and rewards were posted for his capture. He was Arizona's most feared, most wanted criminal, and when the Rangers were organized Burt Mossman was notified that Chacón's arrest was of top priority.[15]

Burt Alvord and Billy Stiles were two criminals who became key figures in Mossman's plans to seize the wary Chacón. Both Alvord and Stiles had been Arizona peace officers before turning outlaw. In 1886, when John Slaughter was elected sheriff of Cochise County, he deputized Burt Alvord. Alvord was dark and husky and bald, even though he was just in his early twenties. He lived in a shack on Tombstone's Tough Nut Street, worked at the OK Corral and Livery Stable, and sometimes drove a stagecoach. Much of his time was spent playing pool or pulling practical jokes. He was handy with his fists and became a noted marksman. Alvord reputedly liked to practice with his revolvers by shooting with his right-hand gun at a string forty feet away; the string suspended a can from a limb, and when he broke the string he tried to drill the can with his left-hand gun before it struck the ground.

Slaughter pinned a badge on this gritty customer, and after "Texas John" left office in 1890, Alvord worked as a peace officer in the Cochise County mining camp of Pearce.[16] Then he moved twenty miles to the north, accepting a constable's badge in Willcox, where he killed "Cowboy Bill" King under suspicious circumstances.

Alvord's cronies were disreputable men such as William Downing and Matt Burts, and he deputized a slender, half-Mexican cowboy from Casa Grande named Billy Stiles. On September 11, 1899, Alvord, Stiles, Downing, and Burts engineered the holdup of a train at Cochise Station, ten miles west of Willcox. A few months later Alvord and Stiles planned, but did not participate in, the train robbery at Fairbank in which Three-Fingered Jack Dunlap and Jeff Milton were shot. With clever detective work Bert Grover, then serving as constable at Pearce, uncovered the gang and succeeded in having Alvord, Stiles, and Downing arrested.

Other members of the gang soon joined Alvord and company in the cells at the big Tombstone courthouse. Within a few days of his incarceration Stiles confessed everything, for which he was allowed to come and go virtually at will. But on April 8, 1900, Stiles shot jailer George Bravin, the bullet ripping through the calf of one leg and taking two toes off the deputy's opposite foot. More than a score of inmates, including William

*When Burt Alvord (standing) was constable of Willcox, he deputized a future partner in crime, Billy Stiles (seated).*

Downing, stayed behind bars, and some of the prisoners carried Bravin to a bed. Stiles released Burt Alvord and Bravo Juan Yoas, and the three fugitives bolted down Fremont Street armed to the teeth. They stole two horses from John Escapule and, doubling up on one mount, headed toward the Dragoon Mountains.[17]

Stiles hid along the Mexican line, darting back and forth across the border and maintaining contact with his 200-pound wife, who lived at their ranch about a dozen miles west of Naco along the San Pedro River.[18] Early in 1901 Stiles was recaptured in Casa Grande by Sheriff Del Lewis of Cochise County, but the slippery outlaw escaped again. From a cave hideout Stiles sent his watch and chain to a friend to sell for the benefit of his wife, saying "they have driven me to the woods again, and I leave my wife very short of means." [19]

Alvord went deeper into Mexico, holing up in the Sierra Madre Mountains of Sonora with other desperadoes. He was known to have operated with Chacón, whose hideout was in the same vicinity. (Later Alvord was quoted as having remarked to Stiles: "We have our necks to save. We must figure out a way to catch Chacón and turn him over to the officers.") [20] Alvord was married, but he was cut off from his wife, who by 1902 was contemplating divorce.

Mossman knew that Alvord, deeply in love, faced a painful marital crisis, and he also had learned the general location of the hiding place of Alvord and Chacón. "Mossman was a first-class detective," observed a close acquaintance, "and he liked it." [21] Gradually, the Ranger captain devised a complex plan to seize *Peludo*.[22]

By January 1902 Mossman had established contact with Billy Stiles, who was earning wages by driving an ore team at the Puertacitas Mine.[23] Mossman asked Stiles to help find Burt Alvord, and through him, Augustín Chacón. Mossman dangled the lure of reward money for Chacón, and he told Stiles he would place him on the Ranger payroll and testify on his behalf in court. Then he authorized Ranger meal money for Stiles from

*When the legendary John Slaughter was sheriff of Cochise County, he pinned a deputy's badge on Burt Alvord.*

January 15 through the end of the month. For four days Stiles ate three meals daily — courtesy of the Ranger fund — in Naco, where he perhaps conferred with Mossman. Stiles traveled to Tucson on January 19, then spent the next couple of days at his home in Casa Grande. Next he made his way to Mexico, spending the last eleven days of the month south of the border at the expense of the Rangers.[24] (Arcus Reddoch, a line rider charged with halting smuggling activities along the Sonora-Arizona border, was convinced that while Stiles was employed by the Rangers, he was active in smuggling Chinese laborers into Arizona.) [25]

By April, probably because of information from Stiles, Mossman felt that he knew the general location of the hideouts of Alvord and Chacón, and he was ready to proceed further with his intricate plans. In late April, Mossman traveled to Tucson to ask for cooperation from his friend, Judge William C. Barnes. Judge Barnes promised Mossman that he would write to Alvord's wife on behalf of the outlaw. Then he gave Mossman a letter to Alvord, stating that he would try to persuade her to drop divorce pro-

*Billy Stiles and his wife Maria. This photo was taken in Naco, Sonora, on March 9, 1902.*

ceedings, provided that the fugitive would reenter the United States and surrender. The judge further promised to use his influence to clear Alvord of train robbery charges, and he gave strong assurance that Mossman was worthy of complete trust.

Letter in hand, Mossman returned to Douglas and secretly contacted Billy Stiles. The Ranger captain informed Stiles about the letter, then told him he needed to be led to Alvord so that he could deliver the proposals. Mossman felt that the half brother of Stiles's wife could direct him to Alvord's lair. He wanted Stiles to take him into Sonora, where the brother lived. Unknown in that part of the country, Mossman intended to pose as a recent jailbreaker from Tucson, and with the tarnished Stiles as a companion he thought his cover would be secure. Leading a packhorse, the two men rode into Mexico.

Mossman had confided his plans to no one but Stiles and Judge Barnes. In outlaw territory, with no friends aware of his whereabouts, Mossman became increasingly edgy. He placed not a shred of trust in Stiles and felt compelled to watch the outlaw vigilantly to avoid a double-cross. By the time the two riders arrived at Torres in Sonora, Mossman had decided he would feel safer without Stiles. He sent Billy back with the packhorse, then rode on to find the half brother of *Señora* Stiles.

Mossman located the man at the village of San José de Pima. After assuring him that he was an escaped criminal who wanted to join Alvord, Mossman was told that Alvord and several other fugitives were holed up at an adobe house in rugged terrain west of San José. The half brother sketched out a crude map and described the hideout. Mossman rode out alone, searching through rough country for two days before finally discovering the outlaw fortress.

On Thursday afternoon, April 24, Mossman rode toward a stout adobe building that commanded a thousand-yard field of fire in every direction. There was a view of any approach for a distance of two miles, and canyons offered escape routes in several directions. As he came closer, Mossman noted that battle shutters barred every door and window, and the thick adobe walls were loopholed. The captain dismounted and walked toward the house, tensely feeling the stare of wary eyes from the loopholes. Burt knocked at the door. An uneasy moment passed before the heavy portal swung open.

Mossman was impressed. He faced "a fine looking man, six feet high, built like a Hercules, . . . with a Winchester in the crook of his arm." The only flaw in his sturdy physique was a bent left arm, which once had been broken and set improperly.[26] Alvord had strong, well-formed features and light eyes, and when he spoke it was without hostility. Mossman boldly announced his identity, then thrust forward the paper he had carried from Tucson. The two men stood in the doorway as Alvord, still cradling his rifle, scanned the letter. As he read, several ruffians gathered behind him. Alvord's features changed expressively as he studied the words, then he folded the letter and jammed it into a pocket. The fugitive turned, told his men it was all right, and ushered the lawman inside. Mossman's gaze swept alertly around the big room, and he noted in a separate room six horses, saddled and ready to gallop toward one of the canyons.

Mossman remembered his empty stomach and announced to Alvord that he was hungry. Obligingly, the outlaw leader went for his horse, led the mount outside, then Alvord and Mossman swung into their saddles and set out on a two-mile ride. They halted at a grass shack in front of a little cave in a canyon wall. Several Mexican women were there, and soon Mossman was devouring a plate of eggs and washing it down with coffee.

By now Alvord had pondered the implications of the letter and was ready to talk. He traded Mossman a fresh mount for his trail-weary animal, then the two men rode into the gathering darkness. For most of the night they stayed in the saddle, discussing every aspect of the situation. Mossman wanted Chacón, and he had to have help from men the wily fugitive trusted. Alvord did not want to betray a fellow desperado, but he was chilled by the thought of divorce. Mossman reminded him that his wife would surely go through with the proceedings if he stayed on the run, and he dangled the bait of amnesty before the lonely fugitive.

Alvord finally allowed that he had seen Chacón a few weeks earlier. As daybreak neared he agreed to Mossman's plan but insisted that he needed Billy Stiles to help him lure Chacón into a trap. Alvord said he would try to set up Chacón, and Mossman wearily headed north to contact Stiles and await word from Alvord.

During the next four months Stiles, inspired by the promise of a share of the Chacón reward money and by Mossman's pledge to intercede with Judge Barnes on his behalf, slipped in and out of the outlaw hideaway. Mossman even placed him briefly on the Ranger roster. On August 8, "William L. Stiles" was paid $55, "July salary in full, Private Ariz. Rangers." [27] At last Alvord and Stiles arranged to lure Chacón to the border with spurious plans to rustle horses from Col. W. C. Greene's ranch near Hereford.

Late in August, Stiles returned from one of his Sonora trips with a note from Alvord to Mossman. The conspiracy called for Mossman and Stiles to encounter Alvord and Chacón at a spring sixteen miles below the border in nine days' time. On the first day of September, Stiles and Mossman rode to the spring but failed to make contact with the outlaws. They crossed the border again, camped in an old adobe stable, then headed toward the spring the following day.

Riding south in the gathering dusk, they suddenly met Alvord and Chacón on the trail. All four men wore holstered revolvers, and a long knife hung from a leather scabbard on the opposite side of Chacón's gunbelt. The four riders dismounted, sat on the ground, and conversed in Spanish. Mossman introduced himself with the story of the Tucson jail escape, but the wary Chacón seemed unconvinced. The quartet nevertheless outlined plans to steal Greene's horses, then Stiles, who supposedly was the only one of the group not wanted by lawmen, rode into Naco for supplies. He returned with bacon, coffee, and sugar, and a supper was soon cooked. After eating, they mounted up and rode toward Naco. The Ranger captain, riding in darkness on foreign soil with a trio of lawbreakers, became certain that Chacón distrusted him and intended to shoot him at the first opportunity.

The night was pitch black and Mossman, understandably suspicious of all three of his companions, maneuvered up behind each of the other riders and gingerly felt of the coats rolled at the cantle of their saddles. Probing for hideout revolvers, he discovered only what seemed to be a pair of handcuffs in Chacón's pocket.

When they reached the side of San José Mountain they made a cold camp, picketing their horses with the saddles still on for a hasty exit. The high desert air turned frigid and everyone donned their coats and curled up to sleep. It became too cold to slumber comfortably, but Mossman had no intention of closing his eyes. He pulled the collar of his heavy coat over his face, then slowly drew his six-gun and, peering between slitted eyes at Chacón, kept a weapon aimed at the dangerous killer through the night.

At last the eastern horizon began to grow lighter, and the men aroused and built a fire. Chacón peeled off his coat and took a turn at cooking breakfast. As he occupied himself, Alvord pulled Mossman aside to whisper that he had provided all the help he could. He murmured that he intended to leave and added a warning for Mossman not to drop his guard against Stiles.

Alvord then called out that he was going for water and would be right back. He mounted up and rode away, with Chacón frowning suspiciously at his retreating figure. Chacón warily refused to let anyone walk behind him, and his eyes glared at Mossman. The three men sat down to breakfast, but Chacón restlessly asked why Alvord had not returned.

When the meal was consumed, Chacón pulled corn husk cigarettes from a pocket and passed them around. The men squatted around the fire, re-rolled their *cigarillos,* and lit up. Mossman felt certain that the notorious *asesino* was on the verge of violence. Purposely, the captain let his cigarette go out, then he stood up, pulled a burning stick from the fire, and relit it. He leaned over and dropped the stick back into the campfire with his right hand. As he straightened up he palmed his revolver; with one smooth movement he thumbed back the hammer and leveled the gun at Chacón. Mossman ordered the still squatting desperado to put up his hands. Chacón's swarthy face registered no trace of emotion as he arrogantly thrust his palms toward Mossman, who directed Stiles to disarm Chacón.

When Stiles lifted Chacón's long knife from its scabbard, he reached around, unbuckled Chacón's gunbelt, and let it drop. Next the Ranger captain commanded Stiles to unbuckle and drop his own gunbelt. Stiles obeyed, then Mossman ordered the two men to step back from their cartridge belts. Cautiously, Mossman moved forward and pushed the guns together with his boot. He told Stiles to handcuff Chacón, then ordered both men to turn their backs. The Ranger backed toward a boulder against which they had leaned their Winchesters. His eyes never left Chacón as he lifted his rifle with his left hand and levered a .30-.40 shell into the chamber. Mossman next had Stiles bring up the horses, but Chacón refused to mount, claiming he could not ride handcuffed. Mossman again threatened to shoot, and Chacón climbed, grumbling and swearing, into his saddle.

Mossman wanted to avoid nearby Mexican Naco, where Chacón might have friends, so he directed Stiles to head toward the border fence about ten miles west of Naco. Stiles led Chacón's horse and Mossman rode in the rear, cradling his Winchester alertly. Chacón tried to halt his mount, but Mossman rode alongside and dropped a rawhide *riata* over the *fugitivo's* head. Grimly, Mossman tightened the noose and warned that if Chacón attempted to leap from his horse he would be dragged across the border.

The trio set out again, Stiles pulling on Chacón's reins and Mossman connected from the rear with the *riata*. They soon reached the barbed wire fence at the border, and Mossman produced a pair of wire cutters. He tossed them to Stiles, who snipped the wires. A short ride brought them to Packard Station, where a train was passing northbound for Benson. Mossman flagged down the train with his hat, and the surprised conductor stopped to let the strange procession aboard.

It was a fifty-mile ride to Benson. Only moments after they chuffed into Benson, a passenger train from the west pulled in carrying Sheriff Jim Parks of Graham County. He was returning home after delivering prisoners to Yuma, and he was delighted to encounter the manacled Chacón. The notorious killer was taken back to Solomonville while word flashed around the territory of his remarkable recapture. Mossman's exploit was a sensation throughout Arizona, but he had hauled in Chacón four days after his Ranger commission expired, and the arrest was made on foreign soil.

Mossman knew a controversy was inevitable, so he deemed it wise to remove himself from the eye of the storm. First he had a debt to pay to Billy Stiles. A grand jury was in session under the direction of Judge Barnes in Tucson. They were considering whether to indict Stiles on charges of mail robbery, and following a quick trip to Solomonville, Mossman appeared on behalf of Billy. The Ranger asserted that he could not have captured Chacón without the continued cooperation of Stiles.

Mossman journeyed on by train to Phoenix, where he reported to the governor. The celebrated ex-Ranger captain next was given free passage on the Santa Fe, and he headed out of the territory, although several stops were made for him to accept the congratulations of friends and admirers. Mossman's final destination was New York City, an entertaining haven remote from the legal and diplomatic machinations necessary to dispose of Chacón.

Back in Arizona, Sheriff Jim Parks returned Chacón to Solomonville

and the courthouse jail from which he had escaped in 1897. Chacón's lawyer tried to have his client released because Mossman had arrested him
without authority in Mexico, but Mossman had no intention of returning
to an Arizona courtroom. The territorial legal system — more pragmatic
than that of the late twentieth century — ignored all technicalities (no one
was present who could support Chacón's claim that he had been abducted
from Mexico) and set Chacón's execution for November 22, 1902.

In the courthouse yard, the gallows still stood that had been erected
for Chacón in 1897. On the appointed date, at 1:00 P.M., the outlaw faced
his fate with bravado before a hundred spectators. Chacón had turned
gray-haired, but before the trap was sprung he looked out over his audi-
ence and announced: "I consider this to be the greatest day of my life." [28]

Alvord and Stiles, despite their roles in Chacón's capture and despite
Mossman's efforts, found no sympathy with federal authorities and
bogged down in court difficulties. Alvord "was wined, dined and lion-
ized" in Naco following the seizure of Chacón, and in December 1902
Cochise County dropped charges against him. But United States Marshal
Myron McCord then served Alvord and Stiles with federal warrants, and
in July 1903 a United States grand jury indicted them for mail robbery.

Confined once again in the Tombstone jail, Alvord and Stiles fell
under a vigorous legal assault and had their guns and saddles impounded
because they were unable to pay court costs. Dismayed at this turn of
events, the two criminal partners began to plan a way out of their di-
lemma.[29]

Burt Mossman often stated that he intended to command the Arizona Rangers for just one year. He nurtured grand ambitions as a rancher
and businessman, and the limited salary of a law enforcement official did
not begin to satisfy his aspirations. Perhaps, too, the criticism aroused
over the August brawl in Bisbee soured Mossman on the role of public official.

A major factor in the decision to resign his Ranger commission was
undoubtedly the resignation in April 1902 of Governor Murphy, his longtime friend and the man who first had involved him with the company.
The new governor was Alexander O. Brodie, but Mossman had no inclination to work under anyone besides Murphy. Mossman announced that
his "private interests in Bisbee . . . needed his undivided attention," and
he tendered his resignation to take effect by the end of August. Of course,
Mossman's dangerous pursuit of Augustín Chacón lasted four days past
his resignation date, but it enabled him to leave office in a spectacular
triumph.[30]

Governor Brodie found an effective successor to the captaincy from
out of his own past. During the Spanish-American War, Brodie served as
a major in the regiment which would become famous as the Rough Riders.
He was wounded in action, won a promotion to lieutenant colonel, and
earned the admiration of Teddy Roosevelt. In 1902 President Roosevelt
appointed Brodie governor of Arizona. Brodie soon picked a Rough Rider

*Solomonville, where Augustín Chacón was hanged in 1902 on a gallows erected for him in 1897.*

comrade, Lt. Thomas H. Rynning, to take charge of the Arizona Rangers.

Thirty-six and recently married, Tom Rynning had been an adventurer since his adolescence. Orphaned by the age of twelve, he worked in a Wisconsin lumber camp for a year, then went to a sister's home in Chicago and spent three years as a stairbuilder. In 1882, at sixteen, Rynning went to Texas, worked as a bullwhacker out of Del Rio, then hired out as a cowboy on the Circle S in the Davis Mountains. He made two trail drives to Dodge City. When news of an Apache outbreak arrived in 1885, Rynning and fifteen other cowboys enlisted in the Eighth Cavalry. Private Rynning was stationed in Arizona during the final Indian troubles. He rode dispatch for General Crook, became acquainted with scout Tom Horn, and was a packer under General Miles. After hostilities ceased, Rynning "laid thousands of dobie bricks" for construction projects at Fort Davis, and carelessly lost his front teeth to the kick of a horse at Fort Meade.[31]

At this stage in his life Rynning stood six feet tall but weighed only 140 pounds. He was a natural sprinter. Track and field competition between companies and posts was a favorite diversion of army units, with considerable betting on the side. Rynning ran the 100-yard dash in fifty-two matches and lost only once, to a black soldier at Fort Robinson. His

best time was ten seconds, and his favorite victory came in 1889 during a
Fourth of July celebration at Deadwood. One hundred yards was marked
off on a downhill slant on Deadwood's Main Street, and Rynning defeated
a champion professional sprinter, Harry Bethune.

At the end of his five-year hitch, early in 1890, Rynning mustered
out, visited his boyhood home in Beloit, Wisconsin, then gravitated to the
Chicago World's Fair. After working there for a while and watching Buf-
falo Bill's Wild West Show, Rynning journeyed to Los Angeles and traveled on to Tucson to become a contractor. He specialized in building railroad bridges for the Southern Pacific. But when the Spanish-American
War erupted, adventure again called and he joined the First Volunteer
Cavalry — soon to be known as the Rough Riders.

Appointed a sergeant, Rynning was promoted to second lieutenant
when the regiment drilled in San Antonio. In Cuba he contracted a fever,
but he still fought at Las Guasimas. When Capt. Bucky O'Neill was
killed, Lieutenant Rynning led the company in the charge up San Juan
Hill. After mustering out, he returned to his contracting activities in Arizona, until Brodie tapped him to lead the Rangers. He received his commission on August 29.[32]

Fall brought a change of headquarters, as well as a new captain, for
the Rangers. Captain Rynning decided that at least one Arizona town was
more "unprotected and exposed" than Bisbee. Of course, no street in the
territory offered more sin per block than mile-long Brewery Gulch, but
impressive brick buildings were going up all over town, and Bisbee, now a
quarter century old, incorporated and began to lose its rough edges.
Twenty miles to the west, however, on the border near Arizona's southeastern corner, a new community was growing which begged for the presence of Rangers.

In the Sulphur Spring Valley, on a flat dusty plain with mountains to
the north and east, the Phelps Dodge Company of Bisbee had located a
prospective townsite in 1900. The town would be christened Douglas,
after James S. Douglas, son of the president of Phelps Dodge, Dr. James
Douglas. It was built to provide a site for a vast new Copper Queen
Smelter. The Phelps Dodge Company had just acquired mines at Pilares
and Nacozari, Mexico, and since loaded ore wagons would have a downhill run into the Sulphur Spring Valley, a new smelter was to be erected in
the vicinity. Alert speculators hurriedly claimed and staked out a townsite
two miles east of the proposed smelter location.

Merchants, saloonkeepers, prostitutes, and gamblers flocked to
Douglas, and when construction workers arrived to begin building the
smelter, they found a cluster of tents and shacks that offered a variety of
goods and services. More substantial buildings soon were under construction, but life in the border town was raw. The streets were so dusty that
planks were laid out to connect boardwalks at major intersections; the
first cafe was built of railroad ties; the only supply of water for the entire
community was a solitary well; and the lone bathtub in Douglas was lo-

*Two former Rough Riders, President Theodore Roosevelt (left) and Governor Alexander Brodie. Shown here with the president at the Grand Canyon in 1903, Brodie appointed another Rough Rider, Tom Rynning, to succeed Ranger Captain Burt Mossman.*

cated in a barber shop. Popular dives lined Tenth and Sixth streets and included the Cattle Exchange, the Waldorf, and the White House. There faro, poker, roulette, and other games of chance were played by eager participants, and painted women danced and sang sentimental hits of the era. Soon the most notorious honky-tonk in Douglas was the Cowboy's Home Saloon, located on unruly Sixth Street.

In 1901 a depot went up on the west side of Douglas to service the El Paso and Southwestern Railroad. The town's main thoroughfare was G Avenue, which ran north and south. The largest mercantile establishment, located on G Avenue, was the Phelps Dodge Company store. Other substantial buildings began to go up nearby, two-story brick structures which contrasted markedly with the low, flat-roofed adobes that sprawled across most of Douglas. Just south of the International Line, a few adobe huts stood around a spring. This little community, Agua Prieta (Dark Water), grew along with Douglas.

Douglas attracted the type of frontier riffraff who always gravitated to new border towns. The saloons, gambling houses, and bordellos mixed in with the old-time excitement of a pioneer community on the grow. An added advantage was the proximity of Douglas to the Mexican border. If peace officers appeared in town, lawbreakers could slip over to Agua Prieta, where sordid pleasures awaited among the *cantinas* and cribs of the *ranchita* on Calle Cinco. Likewise, men wanted by the law in Mexico found Douglas a ready haven. By 1902, two years before the completion of the smelter that gave the town a reason for existence, Douglas was filled with hoodlums from two countries. Traditional frontier vices — drinking, gambling, prostitution, violence — abounded, and tinhorns, felons, smugglers, and assorted other badmen congregated in Douglas.

Captain Rynning, in considering a move of Ranger headquarters to Douglas, felt that he "could operate better from there than from Bisbee up in the Mule Mountains, where Mossman had kept his headquarters. And Douglas itself was the toughest proposition then on the whole American border." [33] Ranger Joe Pearce concurred: "Next to Clifton [a rugged mining town 130 miles north of the border], it was the toughest town in the Territory, the one most in need of cleaning up." [34]

Rynning moved his headquarters into Douglas on October 1, locating the Ranger base on Fifteenth Street, in the south end of town. Ranger headquarters consisted of a two-room adobe, facing north, and a plank corral out back. The front room of the little building served as the office, while the back room was cluttered with bedrolls, guns, saddles, and tack. Even at Ranger headquarters it was unsafe to leave gear out in the corral. There were mangers and feed troughs in the corral, but no barn was erected because a Ranger had no time to retrieve a horse from a barn. Since Rangers stayed ready for emergency calls, at least one horse stood saddled and bridled in the corral twenty-four hours a day. A telephone was installed in the office, and other messages arrived by telegraph or personal callers. From 1903 through 1906 the desk usually was manned by

*At the age of thirty-six, ex-cavalryman Tom Rynning won appointment as the second captain of the Arizona Rangers.*

Sgt. Arthur A. Hopkins, who was enlisted by Rynning to handle clerical duties. Most Rangers held regular field assignments, but four or five men normally were on duty in Douglas, and they bunked in the back room.[35]

Rynning took over the Rangers in September, and within a week Bert Grover and Leonard Page served out their enlistments and chose not to rejoin the force. Only four of the original Rangers remained on duty: Thomas J. Holland, Fred Barefoot, John Campbell, and Henry Gray. Holland reenlisted in September, but he resigned two months later. Just four charter Rangers served out the full year of their initial enlistment, and only Henry Gray stayed on the force longer than a year and ten months. Mossman's original selections proved to be a restless lot, serving brief tenures before moving along to other work. Of course, the pay was not especially good and, as demonstrated by the death of Carlos Tafolla, the work was dangerous.[36]

Because of the frequent turnover, Captain Rynning would be able to staff the Rangers with men of his own choosing. On his second day as captain, Rynning enlisted James T. "Shorty" Holmes at Bisbee. Holmes was a native of Denmark who had become a trombonist in Michigan until he "got something wrong with his lip and had to go to work at something else." Holmes was attracted to Arizona, where he worked for a time as a cowboy. He went on to serve the Rangers for seven years, and was appointed sergeant in 1904.[37]

Later in September, Rynning signed on Frank S. Wheeler and Texan William W. Webb, who had served as a private in Rynning's troop of Rough Riders.[38] Wheeler, a Mississippian, was promoted to sergeant in 1903 and was an outstanding Ranger for seven years. Before the end of 1902, Rynning appointed three more Rangers: Bob Anderson, John Foster, and Bud Bassett. All of Rynning's first selections were former cowboys, and all but one — Webb — served multiple enlistments. Foster, a former peace officer, was appointed sergeant by Rynning.

From October 6 through October 20, a Ranger contingent led by Rynning was on duty in Globe trying to control a strike at the Old Dominion Copper Company. This was an era of violent labor disputes: it was rumored that the miners would flood the main shaft of the Old Dominion at the 12,000-foot level, and mine superintendent Fred Hoar had received a death threat. The Ranger presence helped to curb the situation, and the fortnight at Globe passed largely without incident. In the future, however, there would be far more involvement by the force in the labor troubles of Arizona.[39]

The principal arrest during October occurred in Santa Cruz County. On Tuesday, October 28, two Mexicans, "names unknown," were caught rustling cattle. The stock thieves tried to shoot their way past the Rangers. Pvt. McDonald Robinson's horse was killed, but the Rangers returned fire so vigorously that the rustlers quickly surrendered. The territory

*In October 1902 Rangers controlled a strike by miners unhappy with conditions in Globe's Old Dominion Copper Mine.*

reimbursed Robinson $100 for his horse, but when his enlistment ended in December he left the force.[40]

The Rangers made more than twenty important arrests during November and December. More horse and cattle thieves were seized, along with a New Mexico fugitive and a variety of Arizona lawbreakers. The Ranger record for the first full year had been a good one and would lead to important consequences for the force in 1903.

# 1903: Peacekeeping at Mines and Saloons

*"Arizona now has one of the finest bodies of rangers ever recruited for service on the frontier."* —Governor Alexander O. Brodie

There were few major arrests in the first month of 1903, and Alexander R. "Lonnie" MacDonald was the only recruit enlisted.[1] But in February the Rangers experienced another fatal shootout.

The Cowboy's Home Saloon was the center for drinking, gambling, and dancing in Douglas. It was run by Tom Hudspeth, Walker Bush, and Texas hardcase Lorenzo "Lon" Bass. Late in January, Pvt. William W. Webb, a former Rough Rider, sauntered into the Cowboy's Home and immediately had words with Bass, who resented the presence of Rangers. Captain Rynning later reminisced that Bass and Hudspeth sought him out "and told me they'd kill Webb next time he shoved his nose inside their place."[2]

A couple of weeks later, on Sunday, February 8, the town dives were doing a roaring Sabbath business. When two celebrants created a disturbance in the Cowboy's Home, deputy constable "Long Shorty" Corson sought help from the Rangers. Captain Rynning, accompanied by Webb, Lonnie MacDonald, and William Peterson, went with Corson into the Cowboy's Home. The peace officers quickly arrested the pair of rowdies, to the annoyance of Bass.

That evening some shots went off near the Cowboy's Home. Webb and MacDonald heard the gunfire and hustled over to Sixth Street to investigate. When the two Rangers entered the saloon, Bass sighted them from a rear room where he was dealing monte. He promptly stormed into the main area of the saloon and confronted Webb.[3] Bass angrily ordered Webb off the premises and threatened to "beat the face off him." [4] Some accounts related that Bass produced a revolver and laid open Webb's cheek with a blow from the gunbutt. Other witnesses insisted that Bass had no gun, speculating that one of the saloon girls threw a glass, which struck Webb on the cheekbone.

Surrounded by a hostile, hard-drinking crowd and suffering at least verbal abuse from a man who had previously threatened him, Webb exploded into action. In one rapid motion he whipped his six-gun from its holster, cocked it, and fired pointblank at Bass. The heavy slug spun Bass around, but Webb thumbed his hammer back and fired again. The second round also went true, hurling Bass to the floor.

"Oh, my God!" he gasped as he went down.[5]

Both bullets had torn into Bass's torso, and one apparently struck his heart. He died on the spot.

A few feet away Lonnie MacDonald sagged to the floor, calling out to Webb to shoot anyone else who made a play. A slug had struck MacDonald about waist-high, then ranged upward into his right lung. Apparently, one of Webb's bullets had torn through Bass and hit MacDonald, although some conjecture held that one of the saloon denizens had triggered a round that wounded the stricken Ranger.

Two men now were down, and burned cordite mingled with the thick layers of tobacco smoke in the dim saloon. Only one light illuminated the room, "for those old-time coal-oil lamps would always blow out when there was much shooting any place they was burning." [6]

Captain Rynning and Pvt. Frank Wheeler, patrolling the streets on horseback, heard the shots and spurred toward the Cowboy's Home. Dayton Graham, now serving Douglas as a constable, also hurried to the scene, along with another Ranger or two. Graham, of course, had briefly been sergeant of the Rangers, and it was decided to place Webb into his custody. No jail had yet been built in Douglas, so Graham conveniently put two Rangers in charge of the prisoner.

Quickly gathering information, Rynning found pencil and paper and began sketching the shooting site. He diagramed the saloon, designating the location of the corpse, the bar, gambling tables, and rooms in the rear of the building, and he carefully marked where Webb and Bass had stood when the shots were fired. Later this sketch would prove helpful in exonerating Webb.

Physicians probed unsuccessfully for the slug that struck MacDonald. Douglas had as many hospitals as jails, so the bandaged MacDonald was carried to Ranger headquarters. Rynning's house was located nearby, and his young wife took over the care of MacDonald. Lonnie's fellow Rangers then decided to lend a hand. When Margaret Rynning went over to change his bandages the next morning, the Rangers had just finished preparing breakfast for the convalescent. They had cooked "a big

*W. W. Webb on the outskirts of Douglas. In 1903 Private Webb killed Lon Bass in a Douglas saloon shootout.*

round steak with a lot of greasy spuds and some gravy that a fork could stand up in." Mrs Rynning persuaded them to let her feed him soft-boiled eggs and other light fare, at least until his fever dropped.[7]

Bass, who was a widower, left several orphaned children, and there was considerable remorse — especially among Douglas's large lawless element — against Webb. A coroner's inquest swiftly exonerated the Ranger, but the insistence of several witnesses that Bass had no gun caused Cochise County Sheriff Del Lewis to arrest Webb for murder. Rynning accompanied Webb and Lewis to Tombstone, where the Ranger was held until the end of the week.

On Friday, February 13, Webb was back in Douglas for a preliminary hearing in a crowded little justice of the peace court, the only legal facility which the unincorporated community could boast. A number of eyewitnesses gave testimony unfavorable to Webb, although two men, Alex Gilchrist and James Goode, frankly stated that they were drunk during the shooting and could offer no reliable observations.

On the day of the hearing, a Mexican prostitute quarreled with a man in a back room of the Cowboy's Home Saloon. She threw a coal oil lamp at the man, but he eluded the missile, which crashed into a wall and burst into flames. The ramshackle frame structure, lined inside with cloth and wallpaper, was ablaze within moments.

Fire was an ever-present danger in southwestern communities, and Douglas already had installed fire hydrants. But when equipment was brought to the scene and a hose hooked up, a stream of water could not be produced for twenty minutes. By that time the Cowboy's Home was a loss, along with an adjacent shack which housed a restaurant. The flames had spread to the nearby Copper Belt Theatre, but owner J. O. Phillips darted in and out rescuing boxes of cigars, bottles of liquor, and a cash register. Even his piano was saved, but the building was consumed.[8]

Webb was indicted for murder, and the justice of the peace set his bail at $1,000. Charlie Overlock, a leading businessman sympathetic to the cause of law and order, immediately produced the cash and Webb was set free. During the June session, there was a four-day trial in the ornate courtroom in Tombstone's courthouse. Rynning showed his diagram, and the jury decided that Webb was not guilty. Although there was some criticism of the verdict, Webb had borne no intention of ending up like Carlos Tafolla, and twelve Arizonans endorsed his forthright actions and demonstrated support for the territorial Rangers.[9]

A landmark manifestation of the growing support for the Rangers occurred on March 19, 1903. On that Thursday in Phoenix the Twenty-second Legislative Assembly passed an act doubling the size of the company. Ranger inroads against rustlers and general outlawry had been considerable, although a few county sheriffs and deputies were jealous of the comprehensive authority of the territorial officers. But fourteen men could hardly cover the vast expanse of Arizona. From the start, Ranger efforts had been concentrated on southeastern Arizona, where rustlers thrived along the borders of Mexico and New Mexico. Consequently, numerous counties had been neglected; not a single major arrest had been made in eight counties. It was expected that an expanded Ranger force would permit men to be stationed throughout the territory.

Act No. 64 called for the company "to consist of one captain, one lieutenant, four sergeants and not more than twenty (20) privates." The captain's salary was raised to $175 a month, while the lieutenant would receive $130 monthly. Sergeants' pay was elevated to $110 each month, and that of privates to $100.[10]

Rangers were originally required to provide "a suitable horse, six shooting pistol (army size) and all necessary accoutrements and camp equipage." Now each Ranger also had to procure a pack animal. Captain Rynning thereupon was directed to auction off all pack animals and pack equipment, and forward the proceeds to the territorial treasurer for deposit in the Ranger fund.

Another new provision gave badges to the Rangers. The original concept of the force called for the men to operate in secret, and even though for most Rangers anonymity proved impossible, no badges had been provided. Twenty-five silver badges were handcrafted, each one a five-point ball-tipped star with "ARIZONA RANGERS" engraved in blue on the front. The captain, lieutenant, and sergeants had their ranks engraved on the badges, while privates' badges were numbered. When a Ranger left the force, he was required to turn in his badge to the captain so that it could be reissued to his replacement.[11] Contemporary photographs prove that Rangers, often dressed in surprisingly dapper fashion, proudly displayed their silver stars on vests or coats. When the men worked undercover, however, the badge was "mostly packed in our pocket or pinned inside our riding jacket where it wouldn't be noticed." [12]

Rynning, an administrator by instinct, now had to conduct a major reorganization. He maintained as his second-in-command Sgt. John Foster, now promoted to lieutenant. Jack Campbell, who had been one of Mossman's first recruits, was elevated to sergeant. Ranger records are unclear regarding when other men were promoted to the three remaining sergeant's vacancies. Lonnie MacDonald was made a sergeant, perhaps at this time, and in October Frank S. Wheeler was promoted to sergeant. Harry Wheeler spent less than three months as a private before his elevation to sergeant on October 15.[13] Arthur A. Hopkins, who long would serve as headquarters clerk, enlisted on April 1, but he was not promoted to sergeant until August 1904. William D. Allison enlisted on April 27; he had spent a decade as sheriff of Midland County, and during a two-year stint as a Texas Ranger he had served as first sergeant of Company D under the famous "Border Boss," Capt. John Hughes. Rynning soon appointed Allison first sergeant of the Arizona Rangers.

Now authorized to double his manpower, Rynning began a vigorous recruiting program. Ten new men were signed up in April, five of them on the first day of the month: Hopkins, a former soldier; Sam Henshaw, a cowboy from Texas; David Warford, a forest ranger and ex-Rough Rider; cowboy Clarence L. "Chapo" Beaty; and twenty-eight-year-old Jeff Kidder, a gun-crazy native of South Dakota who would prove to be the most notable of this group of recruits.[14]

Later in April, Rynning enlisted William D. Allison; cowboy James D. Bailey; Tip Stanford and Owen Wilson, cowpunchers from Texas; and "Timberline Bill" Sparks, who had worked cattle. Wilson, Henshaw, and Warford lasted only a few months, discharged "for the good of the service." Otherwise Rynning had picked well. Each of the other seven April recruits served multiple enlistments. Stanford, Sparks, Kidder, Beaty, and Hopkins in time earned promotions to sergeant, and on October 26 Allison was elevated to lieutenant, replacing Foster, who resigned to accept a commission as U.S. deputy marshal. (Foster later rejoined the Rangers, serving until March 15, 1907.) [15]

Before the year ended, Rynning recruited ten more men. They were cowboys and former lawmen, and Charlie Rie had been a blacksmith. Three of them left the force after only a few months, but John J. Brooks, who enlisted in October, was promoted to lieutenant the following April, while a July recruit, ex-cavalryman Harry Wheeler, became the most outstanding Ranger in the history of the force.[16]

Joe Pearce, who signed on in November, later related an unusual en-

*Deputy Sheriff C. H. Farnsworth (left) and Pvt. W. K. Foster. Foster enlisted in 1903 but left the company before the year ended. Farnsworth enlisted in the Rangers in 1905, also serving only a few months. Note Foster's Ranger badge and 1895 Winchester, as well as the double-looped cartridge belts on both lawmen.*

listment experience. Although born in Iowa, Pearce had grown up in northern Arizona. He learned to ride and shoot, became a cowboy, and then a forest ranger in the Black Mesa Forest. But by 1902 he was attracted to the Arizona Rangers: "It sounded like four aces to me." [17] When the sheriffs of Navajo and Apache counties were informed that one Ranger would be appointed from the district, Pearce's name was submitted and he applied to headquarters. Living in Springerville, he received a letter in October from Rynning which reveals the requirements for recruits of that period.

Dear Sir:

In reference to the matter of your application and correspondence with our first sergeant W. T. Allison, where you apply for the appointment of Arizona Ranger, I wish to ask you whether you have the following qualifications. Have you been in the cattle business, and continuously for the last eight years, and have you a working knowledge of the Spanish language? Let me know whether you have such knowledge.

We have made it a practice to enlist only single men, as we have found from our experience that they render the best service, as the men of the service have to abandon their homes and their families and are liable for service in any part of the Territory and thus may be called from their families for as much as six months or a year at a time.

Each man to enlist must provide himself with a horse, saddle, and complete pack outfit, a Colt .45 six-shooter (no less) and a carbine .30-.40 (no less). Can you secure such an outfit? Let me hear from you

*Badges were issued to the Rangers in 1903. When a man resigned he turned in his star; the last Ranger to wear badge number eleven was John McK. Redmond, who enlisted in 1908.*

whether or not you're single, can speak Spanish, are a practiced cowman, and can furnish this complete outfit. If so, we can arrange for you to come to some point and be inspected and enlist you.

Very respectfully
Thomas H. Rynning
Captain Arizona Rangers[18]

Notified of his acceptance in November, Pearce put together the necessary gear, settled his big white hat on his head, and rode south. He made his way through the rugged White and Blue mountains, then headed across the Sulphur Spring Valley toward Douglas. He hit town at dusk, wandered about the dusty streets for a while, and finally entered a gambling hall to try his luck at roulette.

Pearce dropped twenty dollars in a hurry, so he sauntered back outside to find other entertainment. A Mexican circus had crossed the line and set up in Douglas, and Pearce strolled past several of the concessions. Suddenly, he was attracted by a large crowd and the sound of pistol shots.

A Mexican concessionaire and his wife were running a shooting gallery against an adobe wall. Each participant paid twenty *centavos* per shot to plunk away from a distance of twenty-five feet at a string. A Mexican silver dollar with a hole in it was suspended by the string in front of the wall, and if a shooter cut the string and dropped the heavy coin to the ground he was paid one *peso*.

Pearce approached during a lull in the action. *"¡Sombrero blanco!"* called out the concessionaire. *"¿Usted trata, no?"*

"Sure," replied Pearce, "I'll try." He purchased a dozen shots, then readied his Colt. His first bullet severed the string, and the onlookers shouted enthusiastically. His second shot also cut the string, and the concessionaire reluctantly presented two *pesos*.

Pearce missed on his third round, but he continued firing, cutting the target often enough to begin making up his roulette losses. A tall man had moved close, and Pearce turned to him.

"How's that?" crowed Pearce.

"Pretty good," came the noncommital reply.

"Pretty good," snorted Pearce. "Hell, it's damn good."

"I know men who can shoot better and faster. They got to know a lot of other things besides."

"Who are you?" asked Pearce suspiciously.

"I'm Captain Rynning, Arizona Rangers," grinned the tall onlooker. He extended his right hand. "And you're Joe Pearce. Let's have a drink and talk things over." [19]

Pearce signed his enlistment papers at the adobe headquarters building on Monday, November 23. Like all Rangers, Pearce was given a warrant of authority signed by the governor.[20] He soon began to work through Rynning's recruit regimen.

As a combat veteran Rynning was convinced of the necessity of a thorough training program; he exposed the Rangers to a combination of military marksmanship practice and schooling in police techniques that are sound even by today's standards. Rynning stressed to his men that "one live outlaw was worth two dead ones, and even more than that to the reputation of the Rangers." Toward this end, Rynning instructed his men to stroll nonchalantly past a suspect while glancing at his face to make certain of his identity. The Ranger was to walk by his man, then turn to the side, draw a revolver, and order him to raise his hands. "It may seem cowardly," commented Joe Pearce, "but it is an officer's job to arrest a man and not to kill him." [21]

When disarming a man who had his hands raised, Rangers were taught to approach from behind but never to pull his gun from its holster, since that put the weapon closer to a quick grab by the fugitive. The Ranger was trained to reach around from the rear, unbuckle the gunbelt, and let the rig drop around the suspect's feet. The Ranger then would step backward and order the fugitive to move forward. Only when the suspect was several steps away from the gunbelt would the Ranger secure the weapon.

Although most Rangers were old hands with firearms before enlisting, Rynning subjected his men to precise marksmanship instruction. The ex-cavalryman personally drilled his men on techniques of using their weapons. He emphasized that they should always use a revolver when fighting from horseback since that was more accurate than using a rifle during a running gunbattle. Rynning directed his men in firing their six-guns at practice targets while at the gallop.

When it was necessary to use a rifle, the Rangers were trained to slip their boots from the stirrups and slide off the horse's rump, keeping the animal between the lawman and his antagonist for protection. As the Ranger slid from his horse he was to draw his Winchester from its scabbard.

On a training exercise, Rynning and a Ranger would ride out from Douglas and the captain would point to a tree stump or a bush on a hill and snap out an order to dismount. The Ranger would drop to a prone position and immediately open fire, as Rynning shouted out if the rounds were hitting too long or too short. He schooled his men to judge distances by their eyes; the 1895 Winchesters were equipped with adjustable sights, but in an unexpected fight the time used in handling the sight might prove fatal.

Rynning also drilled the Rangers on his version of the term "throwing down": when getting the drop on a malefactor, the Ranger was to draw his revolver as quickly as possible and fully extend his shooting arm toward the target. When *keeping* the drop on a man, the Ranger was taught to be "ready to throw down": since a heavy Colt .45, if pointed at a man for several moments, would cause the arm to waver, the hand and gun would be tilted upward, "ready to throw down" and fire at the first flicker of trouble.

Many of the Rangers purchased their weapons at a gun store in Tombstone which was regarded as "the best in the Territory, and we bought new guns rather than second-hand because we were sure we could trust them." [22] Although the Rangers carried 1895 Winchesters, a few Rough Rider veterans considered the Spanish Mauser a superior weapon. Chapo Beaty supplemented his arsenal with a sawed-off shotgun: "You could shoot it from the hip like an automatic," he explained.[23]

Most Rangers carried their revolvers on the right hip with the butt pointing backward. Some men just shoved the big Colt .45's through their cartridge belts, but most wore holsters greased with skunk oil to make them glossy. A few Rangers had scabbards clipped onto the belt from the inside, worn inside the trousers. Some Rangers tied down their holsters, while others placed their guns on the left side, butt forward for a cross draw. Joe Pearce contended "that a gun packed on the left hip with the butt forward made the quickest draw." But, like most Rangers, he carried his revolver on the right hip, because the gun rode more comfortably that way and because the weapon could most easily be concealed by a coat. In addition to revolvers and rifles, some Rangers carried blackjacks: "We had the blackjack for use in saloon brawls when necessary," said Pearce, "but never used it on the range." [24]

Most Rangers eagerly pursued rustlers, murderers, and other felons, and such arrests created favorable opinion for the company. But one recurring Ranger duty proved unpopular with the men as well as with the Arizona public. Labor strife was widespread, and Arizona's influential mine owners did not hesitate to call for the aid of the Arizona Rangers to

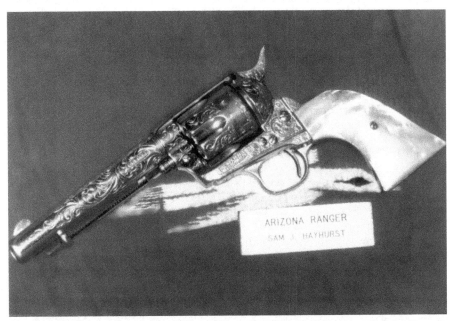

*Sam Hayhurst, a Ranger from 1903 to 1909, carried this beautifully engraved Colt .45.*

quell strikes. On June 1, 1903, an eight-hour-per-day labor law for under-ground miners went into effect in Arizona. Passed by the legislature a few months earlier, the law meant a reduction in daily wages. But miners promptly agitated for an hourly pay hike to make up the difference. La-borers in the Prescott area, for example, received an adjustment ranging from $2.50 for trammers to $3.50 for machinemen and pumpers.[25]

But those men were Americans. Foreigners who worked in Arizona mines found owners less willing to raise their pay. Trouble developed at the big copper mines of the Clifton-Morenci-Metcalf mining district, where Mexican aliens were employed, along with Italian and Slavonian immigrants.[26]

Morenci is surrounded by reddish, green-streaked rock mountains, with cacti, brush, and scrubby trees everywhere. Roads straggle in every direction around the mountains. Clifton, four miles below and east of Morenci, nestles at the floor of a steep canyon, with the San Francisco River (the Frisco) rushing through the middle of town. Metcalf was lo-cated eight miles north of Clifton; today little remains of the Metcalf town-site, which was swallowed by a vast open pit mine.

There was a strong anarchistic element in the labor movement of that time, and a radical group based in Chicago dispatched agitator W. H. Lastaunau to the Clifton-Morenci-Metcalf district. Lastaunau, a fiery Austrian who was called "Mocho" (crippled hand) by the Mexicans and

"Three-Fingered Jack" by the Americans, expertly organized the miners' leaders.[27] A local labor organization was formed, with Abram F. Salcido as president and Frank Colombo as spokesman for the Italians. The workers demanded an eight-hour wage of $2.50, but owners insisted that the low-grade copper content of the district's ore prohibited this increase, which amounted to twenty-five percent an hour. Finally, James Colquhoun of the American Copper Company and James Douglas of the Detroit Copper Company offered a compromise daily wage of $2.25. Lastaunau and the miners, who spent most of their pay in company stores, regarded $2.25 per day as unacceptable. On Monday, June 1, at least 3,000 men in the three camps walked off their jobs.[28]

Lastaunau, Salcido, Colombo, and other leaders held mass meetings in an old lime pit quarried out of a hilltop overlooking Morenci. Liquor flowed freely, and there was angry talk about looting the company stores, dynamiting area railroad bridges, and damaging the mineworks. Sheriff Jim Parks arrived from Solomonville with all available deputies, then deputized company employees and local store clerks, who were scornfully called "counter jumper deputies" by the Rangers.[29] Parks accumulated sixty men, but few were professional peace officers. They armed themselves with "every kind of old gun," for which many had the wrong caliber ammunition.[30]

This unimpressive force seemed incapable of holding at bay thousands of angry miners, so worried mine owners demanded the support of the Rangers. The governor's office gave Captain Rynning his marching orders, directing that the entire company hasten to the scene of strife. Rynning, at headquarters in Douglas, managed to gather seventeen of his men. Charlie Rie, who enlisted the previous day, came along, as did Arthur Hopkins, who normally manned the headquarters desk. Telegrams ordering Rangers to report to the crisis area were fired off to the seven (one vacancy existed in the company) scattered elsewhere around the territory. On Sunday, June 7, Rynning and Lieutenant Foster led their men onto a northbound train. That evening, carrying their rifles and prominently displaying their badges on shirt fronts and vests, the Rangers arrived in Morenci.[31]

Rynning posted his men for guard duty throughout the night. The streets were filled with striking miners and about 500 "tinhorn gamblers, pimps, and other sympathizers." Racist overtones flared. The strikers all were foreign-born, and most frontiersmen were notoriously bigoted. Ranger Bud Bassett observed that the crowd milling in Morenci consisted of "mostly Mexicans, but a lot of Dagoes, Bohunks, and foreigners of different kinds — no whites at all." Despite the tension, there was no trouble that first night.[32]

On Monday the Rangers scouted the area and watched the miners congregate. Large groups of men trooped back and forth between Morenci and the lime pit northeast of town. They carried barrels of highly intoxicating "Dago red" suspended between two poles up to the lime pit, and the strike became festive. By this time Captain Rynning and Sheriff Parks had established a command center at a hotel in Morenci. Rynning rarely emerged from the hotel, however, and Lieutenant Foster assumed a major leadership role. Bud Bassett remembered only one appearance by Rynning, an inspection while riding on a mule. Rynning was always attracted to administrative duties, and Bassett stated flatly that "Johnny Foster took charge of the Rangers in the field." [33]

Monday night was uneventful, but by Tuesday morning, rumor was rife that several hundred striking miners were heading from Metcalf for Morenci, perhaps to loot the big Detroit Copper Company store. Several Rangers in their high-heeled boots hiked up to the West Yankee Mine, where a fire had been built in the boiler. Because of the threat of trouble it was decided to pull the fire, but the Rangers hunkered around the embers since rain was falling and the morning mountain air was chilly.

Lieutenant Foster ordered Bud Bassett and Henry Gray to intercept the Metcalf miners and tell them not to come to Morenci. Bassett and Gray moved out down the trail through Clifton and on toward Metcalf, but they found no one along the route. Later they learned that the Metcalf strikers had left the trail to march up Chase Creek Canyon. Chase Creek empties into the Frisco at Clifton. By the time the Metcalf men headed up Chase Creek toward Morenci, the creek was swollen with water. All the while Lastaunau and company were haranguing the men in the lime pit to a violent pitch. The miners were known to be armed, and the Rangers formed a skirmish line.[34]

At this ominous point, however, nature intervened to avert violence. It started "raining pitchforks," and the torrential downpour broke the ranks of the Metcalf contingent before they reached the pass to Morenci. They clambered to the top off the hills that overlooked Morenci from the north and northwest, seeking shelter from the cloudburst beneath rock overhangs. The Rangers held their positions during the storm, and water overflowed the boottops of Arthur Hopkins.

The deluge broke up the strikers and averted a clash with the Rangers, but the price was a flood which raged through Clifton. Sheets of rain fell in torrents, then walnut-sized hail cascaded down. The roar of rain and hail was interspersed with continued thunder and lightning which reverberated deafeningly off the walls of the narrow canyon. As people sprinted for cover in houses and stores, a wall of water six to eight feet high roared down Chase Creek, carrying houses, wagons, animals, and luckless individuals into the Frisco. Frame buildings were swept away, and debris piled up against other structures, which in their turn were shattered by the accumulated pressure. At least twenty people were drowned, many of them being carried inside their houses into the river, then desperately fighting for life in the raging current before disappearing beneath the water.[35]

The cataclysm ended within a few minutes, but it was an hour before the Frisco receded to its banks. Frantic efforts resulted in the rescue of numerous persons trapped inside wrecked buildings, although most of Clifton's brick structures stood firm on their foundations. Hundreds of citizens, many of them striking miners, began the work of clearing out the wreckage and rebuilding the shattered community. The Rangers again stood guard that night, but everyone in the vicinity now was subdued.

On Wednesday morning, however, the Rangers learned that some of Sheriff Parks's deputies had been disarmed by strikers at the Longfellow Mine, northeast of Morenci. Lieutenant Foster immediately took several Rangers and climbed to a strategic position high above the mine. There was no sign of a problem, so Foster led his men back to Morenci. But armed strikers had overrun the Detroit Copper Company Mill and disarmed several deputies, and a number of mining officials received threats of violence.

The Rangers ate lunch in the Morenci hotel where they headquartered. Then Foster, seeking a way to grasp the initiative, decided that he could end the strike by arresting Lastaunau.

"Bud," said Foster to Bassett, "I guess it's up to us to get Mocho."

"We'll try," replied Bassett.[36]

Carrying their .30-.40 Winchesters, Foster and Bassett set out for the lime pit. Apparently they intended to seize Lastaunau in the midst of his followers, a risky endeavor for two men. But halfway up the trail they saw a man coming down who resembled Lastaunau.

As he approached Foster and Bassett, Lastaunau realized that they were lawmen. "Don't you fellows go any further," warned the labor leader, "or my men will kill you. All I got to do is make a motion."

But without hesitation the two Rangers leveled their Winchesters. "And all I have to do is pull trigger and you're a dead Mexican," snapped Foster with more mettle than ethnic accuracy. "Best thing for you to do is come along with us."

Lastaunau was a hard customer, but Foster's voice crackled with authority and the Winchester muzzles convinced the labor leader to submit to arrest. The three men quickly descended the trail.

"I'd like to make some signs to my men," requested Lastaunau.

The lime pit was a quarter of a mile above their position. "Go ahead," said Foster.

Lastaunau repeatedly waved his arms and called out, but there was no recognition atop the hill.

"They're not paying any attention to you," said Foster. "Come on."

Foster and Bassett marched their captive into Morenci and secured him at the hotel. The Rangers and Parks's deputies then split into squads and combed the town. Several other strike leaders were taken into custody, and a house-by-house search yielded an arsenal of weapons. Bud Bassett recalled that "we gathered more different kinds of shooting irons than I ever heard of before — double barrelled guns, revolvers with four barrels that the barrels revolved, guns with brass barrels, and a lot of other kinds — all of foreign manufacture, and I guess some of them homemade." [37]

The miners were thought to have cached arms in the hills, but the crisis now dissipated. Reinforcements were dispatched to the district in impressive numbers. First to arrive was the entire Arizona National Guard, 230 strong under the command of Col. James H. McClintock, adjutant general of the territory. Then 280 cavalrymen from Forts Grant and Huachuca marched in on Wednesday night. President Roosevelt had granted permission for Acting Governor Isaac Stoddard, who was serving while Governor Brodie was on a business trip to Washington, to use federal troops. More soldiers were en route from Texas, and it was decided that the Rangers could return to their regular duties. [38]

On Thursday, June 11, the Rangers prepared to leave, most of them by train to Douglas. Before they departed, however, the Rangers assembled for photographs. It was one of the few occasions in Ranger history in which the entire company was gathered in one place (only Pvt. Jeff Kidder, a recent recruit who may have been detailed to stay at headquarters, was not present), and at least two poses were photographed by Rex Rice, manager of the local Phelps-Dodge Mercantile. Brandishing their rifles, the Rangers lined up in a single wide rank in an open area in Morenci. Another, less formal pose showed several men kneeling or sitting in front of a standing group. But the photographic etiquette of the day remained inviolate: not a man smiled in either pose. [39]

James Colquhoun immediately praised the Rangers to the newspapers, and Rynning soon received a handsome gold watch from the mining companies. [40] Wisely intimidated by the presence of several hundred militiamen and troopers, the miners agreed to return to work at a $2.25 daily wage. A brief strike at Clifton a month later produced no trouble from the sobered miners.

Lastaunau and seventeen other strike leaders were incarcerated in the Solomonville jail, then were transferred to more secure facilities at Tucson. Lastaunau soon was sentenced to a prison term at Yuma, where as "No. 2029" he proved to be an incorrigible inmate. True to his calling, he organized work strikes and grievance committees among the inmates and was thrown into solitary confinement in the "Snake Den" for his efforts. After an eighty-eight-day stint in solitary, Lastaunau planned what became the last major break at the notorious old prison. On April 28, 1904, he led fourteen other convicts in a rush upon the armory, with a brutal assault upon the superintendent and assistant superintendent in the hope of using the officials as shields. But three of the convicts were shot, and an inmate named W. T. Buck rushed to defend the superintendent and his assistant. Buck caught a blade thrust intended for the warden, but he fought on with a kitchen knife until the rioters were overwhelmed. Buck was pardoned, while Lastaunau went back to the Snake Den. "Mocho" was sentenced to ten additional years behind bars, but he died a couple of

The Rangers at Morenci, left to right: Capt. Tom Rynning, Lt. John Foster, Jack Campbell, W. D. Allison, Fred Barefoot, O. C. Wilson, Frank Wheeler, Oscar Mullen, Oscar Felton, Billy Sparks, Bob Anderson, W. S. Peterson, Bud Bassett, Tip Stanford, W. F. "Tex" Ferguson, Sam Henshaw, Charles Rie, Arthur Hopkins, James D. Bailey, W. W. Webb, Henry Gray, Chapo Beaty, David Warford, J. T. Holmes, Alex MacDonald.

Front row, left to right: W. W. Webb (carrying an older model, tube-fed Winchester), Chapo Beaty, David Warford, J. T. Holmes, Alex MacDonald. Second row: Capt. Tom Rynning, Lt. John Foster, Jack Campbell, W. D. Allison, Fred Barefoot, O. C. Wilson, Frank Wheeler, Oscar Mullen, Oscar Felton, Henry Gray. Standing: Billy Sparks, Bob Anderson, W. S. Peterson, Bud Bassett, Tip Stanford, W. F. "Tex" Ferguson, Sam Henshaw, Charlie Rie, Arthur Hopkins, James D. Bailey.

years later. The prison doctor stated that death was caused by apoplexy, caused by inner rage at his personal confinement.[41]

Although most Arizonans agreed with the forceful suppression of threatened mob violence, there was scattered criticism of the action against the miners.[42] Several Rangers left the force not long after their duty as strikebreakers. These men were throwbacks to an earlier day, adventurers gravitating to America's final frontier. They were cowboys and gunfighters, men who loved horses and wide-open spaces and the thrill of the chase, and their values were those of a vanishing era. They became Rangers because, at least to some degree, they wanted to pursue outlaws and experience the exhilaration of adventure and danger. Standing guard over striking miners who toiled for half the monthly pay of a Ranger held little appeal for such men.

At the end of June, Fred Barefoot, one of Mossman's original recruits, and Sam Henshaw, who had been a Ranger for just two months, resigned. The following month Jack Campbell, another charter Ranger, and J. O. Mullen also resigned; David Warford was discharged. In September, Owen Wilson also was dismissed, while William W. Webb elected not to reenlist.[43] Each of these men was quickly replaced, because many hardy and courageous individuals were eager to pin on the badge of an Arizona Ranger.

Following the breakup of the strike, the Rangers scattered back to their assigned posts of duty. Court was in session in the counties of the territory during June, and the presence of a majority of the Rangers was required in various courtrooms around Arizona. Each summer the company's arrest totals dovetailed as members of the company testified about malefactors rounded up in previous months.

Captain Rynning spent seven weeks during July and August touring Ranger stations across the territory. While in the Aravaipa Valley he personally arrested Ike Clancie, wanted for two murders in Texas. (Later in the year Rynning arrested another accused Texas murderer in the Aravaipa Valley, a rancher from Safford named Tom Bell.) In Pinal County, Rynning observed with satisfaction that the Rangers had performed "particularly effective work" in breaking up cattle rustling.[44]

The company still regularly combed the ranges, ravaged that fall by a severe drought, in search of rustlers. The Rangers worked in close cooperation with the Livestock Sanitary Board and the Arizona Cattle Growers' Association. These organizations willingly assisted the Rangers in enforcing new laws recently passed to curb rustling, and ranchers even loaned Rangers horses on occasion. In December, Pvt. W. K. Foster teamed with a cattle inspector named Pruett to round up seventy-two cattle in the Chiricahua Mountains. Every head of livestock had had its brand altered, and Foster and Pruett threw the culprit responsible behind bars.[45]

The same month, in Cococino County, Pvt. Jim Bailey single-handedly took the field against three rustlers who were "borrowing horses in

*Pvt. Jim Bailey (1903–1906) was one of three Rangers who rode more than 1,000 miles to counter rustlers in northern Arizona.*

the dark of the moon." Bailey scouted the heavily timbered high country around Flagstaff in hopes of catching the rustlers in action. He was riding about twenty-five miles out of Flagstaff when he encountered one of these men in the act of skinning a freshly killed beef. The rustler's partners, a pair of desperadoes named Farrel and Broiller, were not in sight, but Bailey moved in and ordered the man to put up his hands. Staring into a Colt .45 the rustler complied, and Bailey directed him to mount his horse.

Bailey and his prisoner headed in the direction of town along a narrow trail that snaked among rocks and trees. Unknown to Bailey, the other rustlers had seen the arrest. Farrel and Broiller rode up the trail, finally selecting an ambush site not far from Flagstaff. As Bailey and his prisoner wound their way down the trail, suddenly Farrel and Broiller rode into the path about sixty feet away. Their Winchesters were leveled at the Ranger.

"Stop," ordered one of the gunmen. "Turn that man loose. I'll take care of him for a while and you ride on and don't look back or I'll start a young graveyard."

"That will be all right," replied Bailey with deceptive accommodation. "I guess maybe I don't want him anyway. I'll roll one and you fellers can go on." [46]

Bailey pulled cigarette makings from a pocket and began to roll a smoke. Resorting to an old ruse, he looked down the trail and pointed a finger. Although the Winchesters never wavered from Bailey, Farrel and Broiller could not resist cutting their eyes to a distant point downtrail.

Instantly, Bailey sprang into action. Well drilled by Rynning, he jumped off his horse on the left side, keeping the animal between himself and his adversaries. As he cleared the saddle he slid his Winchester out of its scabbard, then instantly cocked and aimed the rifle. Bailey ordered Farrel and Broiller to put their hands up slowly.

It was a standoff, but Farrel and Broiller were exposed atop their mounts, while Bailey was protected by his horse. The rustlers dropped their rifles and raised their hands. Bailey carefully secured their revolvers, then took all three men into Flagstaff. They were tried and eventually convicted of grand larceny. (By 1907 Farrel was free, but Ranger Billy Old arrested him in Flagstaff again on horse rustling charges. He escaped conviction, however, because the owner of the horses had failed to register his brand. When Rangers searched Farrel's pasture they discovered many horses with their throats slit and their brands cut out. In 1908, Pvt. Sam Black arrested Farrel, only to have charges filed against him by the rustler.) [47]

Rangers routinely attended roundups and in effect functioned as livestock inspectors. Such efforts were especially intensive along the Mexican border, allowing Governor Brodie to report happily that their activities on the range caused "a saving to the Territory of the salary of five or six inspectors, as the rangers do this work without extra compensation." The Rangers constantly recovered and returned strayed or stolen cattle, horses, sheep and goats, and Brodie was able to boast that "cattle stealing has been practically wiped out in Arizona." [48]

With their ranks doubled, the Rangers also became increasingly active in local arrest situations. Rynning reported only "principal arrests," rarely recording the frequent misdemeanor arrests made by his men. Local and county officers also readily called on the Rangers for assistance. For example, during a five-day period, June 20-24, three Rangers were dispatched to the mining town of Pearce, assisting local officials in guarding the little stone jail and preserving order throughout a threatening predicament. Rynning informed the governor, "I do not report the arrests made where the Rangers were assisting local officers." [49]

As the Rangers closed 1903 in a flurry of demanding but routine activities, two old acquaintances offered the promise of excitement for 1904. Burt Alvord and Billy Stiles had languished for six months in the Tombstone jail before Alvord finally was sentenced to a term in prison. But Alvord no more intended to go to Yuma than Stiles intended to await his own sentencing. The two longtime partners in crime resorted to a familiar solution: on December 20, 1903, they dug through a wall, clambered down a rope, and helped themselves — along with seventeen other delighted inmates — to freedom.[50]

Arizona was electrified by the escape, as well as by reports that Alvord and Stiles had fled into Mexico and robbed a bullion train. But the Rangers felt that the two could not resist returning occasionally to Arizona. The company organized an alert for the outlaws along the border, hoping that 1904 would bring a final confrontation with Alvord and Stiles.

# 1904: A Few Tarnished Badges

*"Is the reputation of a member of the Ranger force to be made and maintained by the muzzle end of a .45 in the hands of a hot-headed man wearing a star?"* — **Bisbee Review**

On the first day of 1904 the Rangers enlisted another Texan, J. R. Hilburn, who had previous experience as a peace officer. Six days later, Texas cowboy Ross Brooks also signed enlistment papers. But despite the brisk recruiting start, fewer men enlisted in the Rangers in 1904 than in any single year of the company's existence.

Rynning's recruits during the expansion of 1903 proved to be a stable lot, so there were few vacancies in 1904. Only eight men joined the force that year, half of them Texans. Indeed, of the 107 men who served as Arizona Rangers, more came from Texas than from any other province; forty-four men — forty-one percent — hailed from the Lone Star State, and a few had been Texas Rangers. The most notable recruit of 1904, Billy Old, joined in August. A native of Uvalde, Old had served in the Texas Rangers, and in the Arizona force he would rise to lieutenant.

The most important promotion in 1904 occurred in the spring. Lt. William D. Allison resigned, although he soon reenlisted (apparently he tried without success to land the constable's position in Bisbee when Dayton Graham resigned in April),[1] and Sgt. Johnny Brooks was elevated to the Ranger lieutenancy. His promotion became effective on the first of April. Within five months of his elevation, Lieutenant Brooks became blooded as a Ranger. Late in the summer, Brooks and another officer

*More than forty percent of the Arizona Rangers were native Texans, and several had been Texas Rangers. Above is Company B of the Texas Rangers in 1900. The third man from the left is Billy Old, who joined the Arizona Rangers in 1904 and eventually rose to the rank of lieutenant.*                                    (Author's collection)

tried to apprehend Charles Douglas in Bisbee. Douglas, wanted in California and northern Arizona on various charges, vigorously resisted arrest. He "attempted to brain" the officers, whereupon Brooks pulled his six-gun and subdued Douglas with a .45 slug.[2]

The most pressing Ranger business early in 1904 was the capture of former Ranger Billy Stiles and his fellow escapee, Burt Alvord. A few years earlier Stiles, Alvord, and Augustín Chacón had received aid from Henry Wood, who helped run the Ashburn Ranch near the border. For two or three months, while the fugitives camped close to the ranch, Wood would buy food for the outlaws from a store in Crittenden and drop it half a mile from ranch headquarters.

Knowing that Wood once had aided Stiles and Alvord, the Rangers three times rode to his ranch to question him. Sgt. James T. Holmes, Charley Eperson, and Rube Burnett (who perhaps rode with the local law; he was not with the Rangers until July 1905) confronted Wood, who proclaimed his innocence.

"We will make you tell," growled one of the Rangers ominously.

"If you do," replied Wood defiantly, "you will be the first man to make me tell."

The Rangers arbitrarily confiscated and penned an Ashburn steer at Calabasas. Wood went after the steer, tore down the gate, and drove the animal back to the ranch, placing him in a corral with 400 other steers. Holmes, Eperson, and Burnett soon rode up and confronted Wood, who was unarmed and seated on a horse. The officers demanded that the steer be given back to them.

"I will give you nothing," snapped Wood.

The three Rangers drew their guns and methodically shot Wood's mount four times. The horse collapsed, pinning Wood underneath, and the lawmen nonchalantly rode away. Mrs. Ashburn heard the shooting, emerged from the house, and ran to help the fallen Wood. The horse was not yet dead, so she lifted its head and the stricken animal reared and lunged, enabling Wood to crawl free. He asked Mrs. Ashburn to bring him his gun, but she refused.

Continuing their harassment, the Rangers brought Wood before Judge Henry Marsteller. Wood, however, was released, and both the judge and Sheriff Charles Fowler admonished the Rangers for arresting the man without proper cause. Undiscouraged, the trio tailed Wood the next day as he went about his range duties. A few days later someone shot Wood's horse from ambush. The animal was not badly hurt and Wood tried to pursue, but even though he saw no one he was convinced that the bushwhacker was a Ranger. When he was an old man Wood reminisced without remorse that Holmes and Burnett, after leaving the Rangers, were shot to death by cowboys in Nevada, and he fondly recalled that "I whipped Epperson [sic]." [3]

Such surveillance by lawmen along the border soon paid dividends. Tom Rynning heard that the outlaws were headed north toward Naco, and he took Sam Hayhurst on a scout into Mexico. [4] The two Rangers missed their prey, but on Friday, February 19, Sgt. Johnny Brooks received a tip that Alvord and Stiles were at the Young Ranch, located a mile west of Naco, Sonora. Sergeant Brooks, Sheriff Del Lewis, and deputy sheriffs Porter McDonald (who joined the Rangers in 1905), Fred Smith, and A. L. Wilson armed themselves and rode to the ranch at about 8:30 in the evening.

Before reaching the house, the little posse split up. Brooks and Lewis approached from the left side, while the three deputies rode in from the right. Brooks and Lewis arrived first, topping a hill and dimly making out Alvord, Stiles, and a Mexican seated in front of the house and eating supper.

Alvord and Stiles called out hellos, and the two lawmen rode toward them, hallooing back. Stiles rose to his feet and queried, "Is that you, Skeet?" When "Skeet" did not reply, Stiles and his companions leveled their guns.

Brooks and Lewis promptly opened fire. Brooks triggered a shotgun, and two buckshot struck Alvord. Wounded in the thigh and ankle, Alvord toppled to the ground, shouting, "Don't shoot!"

Stiles caught a buckshot in the hand, but he and the Mexican threw a couple of shots at the lawmen, then disappeared into the darkness.

"Don't run off and leave me, Billy," bellowed Alvord, "I am wounded."

Alvord's pleas were ignored. Stiles and the Mexican hustled through high grass to a nearby stream, scrambled down the bank, then splashed across and vanished into the night. When Alvord realized he was being abandoned, he banged away with his revolver in the direction of his retreating confederates. Brooks and Lewis counted the shots, and when Alvord emptied his gun they galloped in and arrested him.[5]

Realizing the impossibility of finding Stiles in the darkness, the posse took Alvord into Naco and jailed him. The next afternoon Rynning and Hayhurst rode in from Mexico and were elated at the news. Line rider Arcus Reddoch also returned from a swing into Mexico and stopped by the jail to see the notorious prisoner. When he came to Alvord's cell Arcus did a double take.

"Were you at Tule Springs the other day?"

"Yes, I saw you there," replied Alvord, "but I didn't know you were an officer, then."

"Well, we are even," said Reddoch. "I didn't know you were Burt Alvord then, either." Three or four days before Alvord's capture, Reddoch had ridden from Arizona's Huachuca Mountains to about five miles into Mexico. When he crossed into Mexico, Arcus pocketed his badge. Trail-weary, he led his pack mule to water, unbridled his saddle horse, then sprawled onto his stomach to drink.

Suddenly a twig popped, and Reddoch looked up to see a swarthy traveler cradling a Winchester. Thinking the armed man was a Mexican, Reddoch spoke to him in Spanish. The man offered no trouble, and Reddoch hurriedly bridled his horse and rode on, not suspecting that he had just seen outlaw Burt Alvord. "No doubt he would have shot me had I been wearing my badge at those springs," the Ranger reflected.[6]

Alvord soon was transported to Yuma to begin serving his two-year term in the territorial prison. Billy Stiles, elusive as ever, remained at large, a $5,000 reward attracting the attention of Rangers, other law officers, and assorted bounty hunters alike.[7]

In the spring of 1904[8] there was an outbreak of rustling in the Sulphur Spring Valley. Calves were disappearing, a few at a time, before they could be branded. The chief suspect was a reprobate named Taylor whose spread was located in the Chiricahua foothills adjacent to the Neil and Hershan Ranch. Taylor reputedly had killed a Mexican to acquire his ranch, and his herd increased at an unnaturally prolific rate. "One of his sons, 'Rat' Taylor, was as bad as his daddy," recalled Rynning, "always selling blue meat — the flesh of unweaned calves — round the streets of Douglas." [9]

Taylor had been arrested before, but the calves wore his brand and he was always released. Ranchers in the district complained to Ranger headquarters, and Rynning decided to handle the case personally. He hatched a unique plan to identify the rustler. Accompanied by Johnny Brooks, Rynning rode north out of Douglas into the Sulphur Spring Valley. It was spring and the calf crop was coming in. When they encountered Billy Neil's cattle, the two Rangers roped thirteen red-and-white and black-and-white calves, "well flesh-marked, not weaned and lacking about 30 days before the spring round-up." It was Brooks's idea to cut out thirteen calves — an unlucky number for the thief.[10]

Rynning recently had been in Mexico and had returned with a tobacco sack full of Mexican five-*centavo* pieces minted in 1885. The Rangers opened their pocketknives, made a small slit in the neck of each calf, within the folds beneath the chin, and slipped a coin into each slit. Then they moved the baited calves and their mamas to range nearer Taylor's spread.

A few months later, just before roundup time, Rynning returned to the valley with some of his men. Scouring the Taylor range and feeling the gullets of red-and-white and black-and-white calves, the Rangers managed to find eleven of the baited animals. Rynning sent a man into Tombstone for a four-mule truck, and the calves were taken into town and penned in a corral near the courthouse. Rynning and Brooks rode to Taylor's house and arrested the rustler, who laughed off the Rangers' efforts, crowing that the only brands on the calves were his.

When court convened in Tombstone, the foreman of Taylor's jury was a cattleman named Jacklin, owner of the 7D Ranch. Rynning told the judge about the coins, and the participants left the ornate courtroom and trooped out to the corral. At the corral Rynning was called as a witness. He testified how he and Brooks had planted the Mexican coins, dated 1885, in the calves on Neil and Hershan land. Then Rynning and Brooks handed their pocketknives to the bailiff, who presented them to the jury. With an experienced hand Jacklin deftly opened a calf's neck and produced the telltale coin.

A deflated Taylor pled guilty. Prior to his sentencing, however, an out-of-court agreement was made. The Neil and Hershan Ranch had offered Taylor $16,000 for his land and cattle, considerably less than the true value. The ranchers felt that such a bargain would return to them their rustling losses, and it would remove the likelihood of Taylor's sons continuing the thefts. The ranchers proposed to arrange for Taylor's release, on condition that he and his family leave Arizona. Taylor, of course, agreed, but he insisted that the Rangers bring him the $16,000. Rynning and Pvt. Sam Hayhurst rode out with the money, then hurried the Taylors on their way. It was a satisfying moment for Rynning: "So we got rid of the worst nest of cattle thieves on that range." [11]

Arizona felons frequently took advantage of the nearby border of Mexico. Throughout their existence the Rangers maintained a close cooperation with Mexican officials. "In those days we could go into Mexico if on a hot trail," said Sam Hayhurst, "if we did not stay too long, without

*Burt Alvord passed through this gate to begin a two-year term in Yuma Territorial Prison.*

permission." [12] In addition to the "hot trail" arrangement, on numerous occasions one or more Rangers entered Mexico for extended periods of time while searching for fugitives. Customarily, Rangers would request a leave of absence from the company so that the technicalities of Mexican law would be satisfied. Often the Rangers helped Col. Emilio Kosterlitzky's hard-riding *Rurales* patrol the border for rustlers, smugglers, or gunrunners, whose activities frequently became too large-scale for the customs officials and line riders employed by the United States and Mexican governments. Kosterlitzky and his men, in turn, had aided the Rangers in rounding up fugitives from Arizona. The Rangers had received similar cooperation from Capt. Quintano Molina of the Cananea *Gendarmaria* and from the chiefs of police of Cananea, Agua Prieta, and Naco and Nogales, Sonora. Governor Brodie happily reported in 1904: "The most cordial relations exist between the Mexican authorities and the rangers. They have at all times assisted and cooperated in the following and apprehension of fugitives from justice and the recovery and return of stolen property back to this country." [13] In general the Rangers realized, as Joe Pearce put it, "that when our business took us south of the border we weren't hampered by any legal and political technicalities." [14]

In 1904 two railroad contractors were repairing a stretch of track on the El Paso and Southwestern about nine miles west of Bisbee. Several of their best mules were stolen, and they telephoned Ranger headquarters, stating that they thought Mexicans or Yaqui Indians had stolen the animals and driven them into Sonora.

*Col. Emilio Kosterlitzky (on white horse at far left) and a column of his hard-riding*
Rurales.

Tom Rynning called in Joe Pearce and four other Rangers who happened to be in Douglas. After outlining the situation, Rynning added:
"You boys don't look at the border when you cross it. It's not there." [15]

The five Rangers saddled up and rode to Bisbee, then continued on
to the grading camp, where they learned that two Yaqui muleskinners had
disappeared without collecting their wages. When the Rangers picked up
the trail they noted the tracks of about a dozen mules and perhaps fifteen
horses. They surmised that the two muleskinners had worked with a number of confederates who met them below the border.

The Rangers followed the trail toward Fronteras, a Sonoran town
nearly forty miles south of the international line. They rode from the contracting camp to Fronteras by nightfall. After buying jerky to eat on the
trail and corn for their horses, the Rangers went to find a meal. They encountered twenty-five *Rurales*, who learned why the Rangers were in Sonora and wanted to come along.

The Rangers needed their help but had been cautioned by Rynning
to avoid becoming involved in a killing. *Rurales* were known to be fond of
directing fugitives to dig their own graves, then to summarily execute
them by rifle fire. The Rangers talked the *Rurales* into detailing just five
men to help with the pursuit: five *Rurales* would be welcome reinforcements against fifteen Yaqui rustlers, but the Rangers should be able to
prevent an execution.

The next morning the pursuers picked up the trail with little diffi-

culty. By afternoon the tracks were so fresh that the pursuit party slowed
up to avoid ambush. About four o'clock they topped a rise and viewed a
draw below that led south. A stream wound through the middle of the
draw, which was thick with mesquite and brush and cottonwood trees.
The Rangers led the way to a higher position, from which the Yaquis
could be seen encamped in a small glade. A fire had been built beside a
cottonwood trunk, and the Yaquis were roasting strips of meat from a
freshly slaughtered beef on green willow sticks. Their horses, still saddled
and bridled, were tied to the brush several yards from the fire, while the
stolen mules grazed some distance away.

The Rangers, uphill and at a range of about 300 yards, took their
Winchesters and crawled to firing positions. Pearce and his companions,
aware of the *Rurales's* reputations as poor marksmen from long range, intended to shoot into the campfire and set the rustlers to flight with no fa-
talities. Still mounted, the *Rurales* in Spanish snapped out their desire to
charge into the camp, but the Rangers strongly asserted their plan. After
a "ticklish" moment, the *Rurales* reluctantly dismounted and crawled into
position.

The Rangers opened up, blasting the campfire apart. Hot coals and
ashes flew in every direction, and smoke billowed thickly. As the rustlers
bolted for their horses, the *Rurales* began to bang away, but their bullets
fell woefully short. "We could see them pop up dust fifty yards in front of
the Yaqui camp," reported Pearce scornfully.[16]

Under fire, the Yaquis did not take the delay necessary to untie their
reins. Whipping out their knives, they slashed through leather and rope,
vaulted onto their horses, and spurred away through the undergrowth.

The officers mounted up and guided their horses down to the hastily
deserted camp. Pieces of reins and halter ropes dangled from bushes, and
ashes and meat were scattered about. One horse, still bridled but unsad-
dled, was found nearby. Apparently the rider had uncinched his saddle
after tying his mount, then when trying to escape he had lost the saddle
and sprinted away on foot. The *Rurales* dismounted to retrieve the saddle,
then took possession of the horse. While they scrambled for the saddle,
Pearce slipped off the bridle as "the only souvenir we took back." The
headstall was studded with silver buttons, and the bridle bit was laden
with silver conchos.

The stolen mules were rounded up, then driven northward. At Fronteras the *Rurales* took the saddle horse and parted amicably from the
Rangers. The Rangers reentered the United States at Naco, then returned
the mules to the contracting camp before riding east to Douglas.

In June 1904 Captain Rynning ordered Joe Pearce on an assignment
to his old stomping grounds. Directed to receive further instructions in
Phoenix, Pearce used his railroad pass and loaded his saddle, bridle,
spurs, and Winchester aboard a train bound from Douglas to the capital.
In Phoenix he stopped first at the office of H. Harrison, secretary of the
Livestock Sanitary Board. Harrison reminded the Ranger that because of

drought conditions on the range, ranchers in the Bloody Basin and the Tonto Basin were not intending to round up their weak cattle for calf branding. The situation was ideal for stock thieves, but it was hoped that Pearce's familiarity with the country would help him catch a rustler or two and scare off the others.

Pearce next went to the capitol, where he had an appointment with the governor. Brodie offered the Ranger a cigar, then outlined a threatening situation between cattlemen and sheepherders along the Verde River. Violence was likely when the sheep were moved north to summer ranges, and Brodie wanted Pearce to work out a peaceful agreement between the rivals for grazing privileges.

Pearce took his gear to a northbound train, passed through desert valleys studded with saguro cacti, and debarked at the mountain mining town of Prescott. After an overnight stay he bought a stagecoach ticket for Camp Verde. The Verde River winds through a deep, breathtakingly beautiful valley, and Camp Verde stands beside the rippling blue water. At that time several old cavalry buildings dominated the east end of town. In Camp Verde, Pearce encountered Constable Frank Smith, who related details of a band of mule thieves. The stolen animals were sold in mining towns, where mules were greatly coveted for underground work.

The two officers decided to pursue the mule rustlers together, and Smith located a horse for Pearce to throw his saddle on. They rode downriver along the cottonwood-lined riverbank and soon found the tracks of several sharp-shod mules and a lone rustler. The mules were driven through rough country, but the lawmen managed to follow the cold trail.

Pearce and Smith camped out twice in the brisk mountain air, but on the third day at sundown the trail led into a sheep camp. Four men were clustered around the campfire, and the officers had no idea who were rustlers and who were sheepherders. Pearce briefed Smith on Ranger techniques in such a situation, and the officers nudged their horses toward the camp.

As they rode in, one of the men in camp amiably invited them to get down. Range etiquette demanded that the newcomers should dismount promptly, "for to stay on your horse means you haven't come on friendly business." Keeping their mounts between themselves and their hosts, the officers obediently swung out of their saddles — then threw down across the backs of their horses. The four men around the campfire were caught completely off guard and were easily disarmed.

Two of the men wore revolvers, and the officers handcuffed them, assuming that the men without sidearms were sheepherders who innocently had offered chuck to passersby. Smith agreed to guard the camp through the night, while Pearce snapped a pair of cuffs to the ankle of one rustler and to his own ankle. "Of all the nights I've slept that was the worst, the most uncomfortable," reminisced Pearce, "with that metal eating into my flesh every time my sleeping partner moved or stirred about." The next day the lawmen took their two prisoners to Camp Verde, where a justice of the peace placed each rustler under a $500 bond.

Now Pearce rode on alone to the ranch of John and Pete Latourette, a big cattle spread along the Verde River. John Latourette was there, and when Pearce introduced himself as a Ranger, the florid-faced rancher happily pumped his hand up and down. Latourette confirmed that he was losing cattle, and the two men held a planning session. It was decided that Pearce would pocket his badge and play the role of a newly hired cowboy, but the foreman would be instructed to let him ride around the range at will.

So that he would look like an innocent waddy, Pearce locked up his Winchester and gunbelt at the ranch, concealing his Colt in his outfit. Latourette provided the Ranger with a good string of horses. For the first few days Pearce worked closely with his fellow cowboys, but then he began to scout the nearby Bloody Basin country with a packhorse in tow. Soon he extended his explorations into the rolling grasslands beneath the Tonto Rim, where through a spyglass he saw a rider driving a calf ahead of him. The calf was on a rope, and the rider, horse, and calf quickly disappeared.

Pearce loped his mount to where the man had vanished. Suddenly, ahead of him appeared a grassy little valley, with rim rock forming a natural fence around the top. Where the rim rock dropped too low, natural timber posts and wire filled in the gaps. There was a crude gate tied shut with rawhide. Since the rider was nowhere in sight, Pearce untied the gate and entered the pasture.

There were thirty or forty calves, many of them dogies (already weaned), but also a lot of "wet stuff" (calves still in the process of being weaned). Several displayed skinned legs where they had been hobbled on one side so that they could not keep up with their mamas; such calves thus were weaned on the range and then stolen. Although a number of the calves were unbranded, others had been freshly branded.

Pearce rode among the calves, checking the brands against his brand book. Several calves had been "sleepered," or marked with brands not on record in Arizona Territory. Some of the calves, however, showed the brand of Andy Longfellow. Although there was no conclusive evidence that Longfellow had done anything illegal, Pearce was convinced that the man who had penned the calves was a rustler.

Pearce began a thirty-mile, cross-country ride to Pleasant Valley, where the nearest justice of the peace, Bob Samuels, was headquartered. Throughout the journey Pearce mulled over the problem of how to prove theft. When he found Judge Samuels he explained his plan. The judge located two experienced cattlemen to assist Pearce, then word was spread among area ranchers that stolen calves were being driven to Youngtown, and that cows without calves should be brought to the corral there.

Pearce and his two associates rode to the secluded pasture and drove the calves into town. By the time they arrived, a number of ranchers had come to the corral, many with cows. When the calves were penned, several stolen animals were immediately recognized by small ranchers who knew every head of stock in their herds. Recognition, however, was insufficient proof in a court of law.

Pearce then released the calves, and five unweaned animals went bawling straight to their mamas. The brands on calves and mamas did not match, and with this solid proof of theft Judge Samuels drew up a warrant of arrest for Andy Longfellow.

The next morning Pearce took the warrant, packed into his bedroll enough bread, jerky, coffee, and salt to last for several days, then rode back to the little rock-rimmed valley. There he found a cold trail that led north, upward toward the Tonto Rim. Pearce and his mount clambered through rocky, broken country so heavily timbered that the rustler's trail often was covered by fallen pine needles. As he picked out the gapped trail, Pearce concluded that Longfellow had decided to hide in the high country until it seemed safer to drive out the stolen calves.

On his third day out, at mid-morning, Pearce sighted the smoke of a campfire. Proceeding cautiously, the Ranger soon sighted his prey, stooped over a fire and shaking a frying pan filled with bacon. The man had a revolver belted around his waist, but his rifle was leaning against a tree several feet away. Pearce rode innocently into the camp, and after greetings were exchanged he was offered breakfast.

"Don't mind if I do," said Pearce. "You Andy Longfellow?"

Longfellow stiffened and placed his frying pan on the ground, but Pearce flashed his Ranger badge.

"I've got a warrant for your arrest on a charge of cattle rustling," Pearce said calmly. "You better be careful and not make any blunders with shooting irons."

"I'll go with you."

Pearce took Longfellow's six-gun, then levered the shells from the rustler's Winchester. The Ranger and his docile prisoner next ate breakfast. They rode toward Youngtown, camped out one night, then Pearce turned Longfellow over to Judge Samuels. Longfellow, who was revealed to be a rustler from Texas with a string of aliases, was placed under bond. Pearce later heard that Longfellow, unrepentant, stole a yearling from his brother and was shot to death by his touchy sibling.[17]

The apprehension of Longfellow and of the two men in the sheep camp were typical of Ranger activities. Rynning's report to the governor at the end of the fiscal year, June 31, 1904, stated that the Rangers had made 453 arrests during the twelve-month period. Five men arrested were charged with murder, 155 with felonies, and 293 with misdemeanors.

The captain recommended that twenty-six Krag-Jorgensen carbines be issued to the Rangers. The territorial militia already was armed with the bolt-action Krag-Jorgensens, which were military issue until 1904, when the 1903 Springfield became the army's rifle. But there is no record that the Rangers switched weapons, and photographs show the Rangers continuing to carry their 1895 Winchesters.

Rynning reported that during 1903–1904 the company rode a total of 10,140 miles horseback, an average of 390 miles per man each month. Governor Brodie praised these strenuous efforts: "The moral effect of having the rangers in the Territory, and by their constant appearance on the ranges, in the mountains, and patrolling the Mexican line has been most effective in keeping down crime." [18]

One example of the far-ranging missions of the force occurred when Chapo Beaty, Jim Bailey, and George Devilbiss were sent against rustlers at the request of ranchers north of the Colorado River. When the trio of Rangers rode out of Flagstaff, they knew they could not cross the Colorado. So they veered west through Mohave County, followed the Virgin River through the corner of Nevada, crossed briefly into Utah, and finally reached the trouble area. They had ridden 1,000 miles to counter rustling activities about fifty miles north of Flagstaff.[19]

During this period, one Ranger gained some notoriety for himself and in the process almost blemished the reputation of the organization. Jeff Kidder had the temperament and skills of a western gunfighter, but he was born almost too late to inhabit an untamed frontier. Kidder found an outlet for his lethal inclinations as a member of the Arizona Rangers, and he enforced the territory's laws with almost bullying zeal.

Jefferson Parish Kidder was born on November 15, 1875, at his father's home four miles north of Vermillion, South Dakota. Growing up in Dakota Territory, Jeff handled firearms throughout his boyhood. His father was postmaster in Vermillion, and when Jeff was given a clerk's job he spent most of his pay on ammunition. He practiced incessantly with a Colt .45 and became a crack shot with either hand.

Jeff graduated from Vermillion High School, an unusual achievement for adolescent boys in the West of the 1890s. He still wanted to wield a gun in the time-honored western manner, but the frontier was fast disappearing. Then, in 1901, his father moved to California to improve his health. Jeff, in his twenty-sixth year and still single, went to Arizona, seeking adventure in one of the few areas that still held the promise of primitive excitement. He worked as a miner, then found employment in Nogales as a peace officer.[20]

Jeff was one of the first applicants accepted during the expansion of 1903; he enlisted as a private on the first day of April. Tall and strongly built, with light hair and gray eyes, Kidder was unconcerned about his appearance. He usually wore old black boots and shabby range clothes, but the revolver belted around his waist was always cleaned and oiled. He never stopped practicing with his guns, and he was widely regarded as one of the fastest draws in the Southwest. There were many fine shots with the Rangers, but only Harry Wheeler was considered Kidder's equal as a marksman.[21]

Although Kidder had likeable qualities, even his friends agreed that "he did not have a bit of sense when he was drunk." [22] After a few drinks

*Young Jeff Kidder upon his graduation from Vermillion High School in North Dakota.*

he became overbearing, even pugnacious, and he was inclined to rough up Mexicans when in his cups.

Veteran lawman and gunfighter Jeff Milton was antagonized by Kidder's manner. The two Jeffs had trouble on at least one occasion in Nogales. They were eating together in Pete Cazobon's restaurant when Kidder began cursing. Milton, solicitous of the women and children present in the room, sternly upbraided Kidder. According to the respected historian J. Evetts Haley, Milton's friend and biographer, Kidder angrily whipped out his six-gun. But Milton talked him into holstering his weapon and stepping outside to settle their differences. Once in the street, Milton hotly insulted Kidder, "hoping he would draw on equal terms." But Kidder refused to fight and slunk away in cowardly fashion. Line rider Arcus Reddoch, who knew both Jeffs, described a somewhat different version of the incident, stating that "Jeff Kidder called Jeff Milton's hand and told him plenty." Reddoch theorized that "it was a good thing Milton had Kidder covered, because on an even draw, Kidder would have beat him, everyone said. . . ." [23]

Kidder's first duty station as a Ranger was in Nogales. When the Rangers were searching for Burt Alvord and Billy Stiles early in 1904, Kidder rode alone into the Tule Springs country, but the outlaws at that point were hiding at Monkey Springs. [24]

Because of his heavy-handedness with anyone he considered a troublemaker, Kidder tarnished the Ranger image on more than one occasion. The first such problem occurred in Bisbee on Tuesday evening, July 5, 1904. It is not known if Kidder had been drinking, but he obviously was in a bellicose mood. As he walked around the streets, Colt .45 at his hip and five-pointed star pinned on his chest, he encountered men named Fagan and Graham who somehow gave offense. Kidder manhandled both men, then proceeded belligerently up Main Street. On the sidewalk in front of the Turf Saloon stood two companions, one of them a young miner named Radebush.

The "Desert Rats," a group of Arizona old-timers who gathered in the Pinal Mountains on June 7, 1936. Peace officer Jeff Milton (third from left, rear) was one of
many men who had clashed with Jeff Kidder. Judge Will C. Barnes (far right) was
an early supporter of the Rangers and a longtime friend of Burt Mossman. This photo
was snapped by Mrs. Jeff Milton.

"You'll have to get off the streets here," growled Kidder, shoving
Radebush out of the way.

"Why," protested Radebush, "what's the matter with our standing
here?"

Incensed at a hint of defiance, Kidder executed his lightning draw
and clubbed Radebush in the face with the butt of his revolver. Knocked
senseless, Radebush collapsed, cutting his head when he fell onto the cement sidewalk. A large crowd angrily swarmed to the scene, and there
were shouts to "get a rope," "hang him," and "string him up." But other
officers hustled Kidder away before a lynch mob could organize.

Bisbee, of course, had seen similar problems erupt between citizens
and Rangers, and there was a furor of criticism. "Who sent this Ranger in
here with his pistol to beat up men on the streets of Bisbee?" Kidder was
excoriated as "a thug, bully and butcher." [25]

On Wednesday Kidder was arraigned in Bisbee on two charges of assault and one of assault with a deadly weapon. Because of hard feelings
throughout the town, Kidder requested a change of venue for the first two
cases, which Judge McDonald granted to Tombstone. The Radebush case
was bound over to the grand jury. The citizenry continued to complain,
while Rangers in Bisbee apologized on behalf of the force. [26]

Two weeks later in Tombstone, Kidder was convicted by a jury of assault and battery on Fagan. The judge fined Jeff fifty dollars. The Graham case was dismissed, however, because witnesses for the prosecution
failed to appear. The Radebush case dragged on into 1905; Kidder's attor-

*Bisbee's Main Street. The Bank of Bisbee is at right, while the three-story post office looms at left.*

neys managed to have the trial transferred to Pima County, and it ended inconclusively.[27]

Jeff Kidder retained his Ranger commission. He was an efficient officer with genuine — if sometimes contentious — enthusiasm for his work. He continued to stalk lawbreakers, he never stopped practicing with his guns, and in the future he would repeatedly become involved in violence and controversy.

Not infrequently during the history of the company, an arrest victim filed charges against the Ranger who had taken him into custody. In 1904, for example, Henry Gray was charged with aggravated assault. On October 13 the veteran Ranger appeared in the Tucson court of Justice Culver, who set a bond of $500. Gray put up his bond, then reappeared five days later to hear a grand jury indict him. The next day, October 19, he was arraigned and pleaded not guilty. His case was set for Monday, October 31, but perhaps the matter was settled, because no further mention of Gray's case may be found in the court news.[28]

Each time a Ranger was brought up on charges, the captain or lieutenant appeared to provide supplementary information and character witness. This meant, of course, that valuable time was spent in court by ranking officers defending Ranger methods, and the defending Ranger was out of the field for a substantial period of time. Henry Gray spent much of October 1904 in court, and when he returned to duty he must have been somewhat disgruntled, like other Rangers before and after him, at the legal harassment instigated by the criminal element.

*Harry Wheeler engaged in his first shootout as a Ranger while foiling a Tucson saloon holdup in 1904.*
(Courtesy Arizona State Archives)

Perhaps the most dedicated and lethal of Arizona's Rangers was Harry Cornwall Wheeler. The son of Col. William B. Wheeler of the United States Army, he was born on July 23, 1875, in Jacksonville, Florida. Harry was reared on a succession of frontier military posts in an atmosphere of patriotism and discipline. During his boyhood he was taught to shoot, and he became an exceptional marksman, expert with rifle or pistol. Reputedly he was sent to a military school but was turned down because of the height requirement, which was merely five feet (eventually Harry Wheeler attained a stature of five-foot-six-and-one-half). In 1897 he enlisted in the First Cavalry at Fort Sill, Oklahoma, and the next year he married and soon became the father of a son. Wheeler reenlisted in 1900, earned promotion to sergeant, and was transferred with his outfit to Fort Grant, Arizona. Sergeant Wheeler was given a medical discharge in 1902, having suffered internal injuries when kicked by a horse. Although certified as disabled, Wheeler waived admission to a soldiers' home. He worked as a laborer in Willcox and as a miner in Tombstone, then seized an opportunity which would give his life purpose and notoriety. On July 6, 1903, his application was accepted by the Arizona Rangers.[29]

An iron-willed young man with a stern devotion to duty, Private Wheeler quickly impressed his superiors. On an early personnel report, Captain Rynning wrote of Wheeler: "Excellent service, honest and faithful — gallant, active & intelligent, an officer of great ability." Just four months after enlisting, the twenty-seven-year-old Wheeler was promoted to sergeant.[30]

Though stationed in Willcox, Sergeant Wheeler had official business in Tucson on October 22, 1904. The following evening he was in Wanda's Restaurant on Congress Street when two holdup men approached the nearby Palace Saloon. One man positioned himself across the street as a lookout, while Joe Bostwick crept around to the rear of the saloon.[31]

Inside the Palace four regulars were on duty: night bartender Decker; Lincoln, the craps dealer; Johnson, the roulette dealer; and a black porter.

Half an hour before midnight, the only customers were "Policy Sam" Meadows, E. O. Smith, miner Matt Fayson, and carpenter M. D. Beede.

Bostwick slipped through the rear door of the Palace. He wore a slouch hat, a pair of overalls, and a long, faded coat. He brandished a long-barreled .45 and his face was covered with a red bandana, complete with eyeholes.

"Hands up!" he shouted.

Beede instantly darted out the front door, but everyone else was riveted to attention by the masked man and his gun. "Throw up your hands," Bostwick repeated nervously, "and march into that side room."

As the men moved toward the side room, the jittery bandit snapped, "Hold 'em up higher — hold up your digits."

Bartender Decker, with surprising jocularity, held up one finger, then another, and asked if that was enough. "Get a move on," growled the masked desperado, and Decker raised his hands and followed the others. Bostwick edged toward the craps table, where money was scattered near the dice.

Outside on Congress Street, the fleeing Beede encountered Sergeant Wheeler, who had just emerged from Wanda's Restaurant. Beede blurted out that a robbery was in progress, and when Wheeler turned toward the saloon he added: "Don't go in there — there is a holdup going on!"

"All right," calmly replied the diminutive Ranger, "that's what I'm here for."

Wheeler pulled his single-action Colt .45 and stalked to the front door of the saloon. Bostwick spotted him and whirled to fire his revolver, but Wheeler triggered the first shot. The heavy .45 slug grazed Bostwick's forehead above the right eye. Bostwick fired wildly, then Wheeler drilled him in the right side of the chest. The stricken bandit groaned and collapsed to the floor.

When the shooting started, the lookout man fired a round at Wheeler. The bullet imbedded itself harmlessly in the leg of the roulette table, and in the excitement of the fight Wheeler did not until later realize the source of the fourth shot.

The men in the saloon crowded around the fallen thief, and a physician named Olcott was summoned to the scene. Dr. Olcott ordered the wounded man to be carried to the local hospital.

Wheeler telegraphed news of the shooting to Captain Rynning in Benson, and Rynning caught the first train to Tucson. The day after the shooting it was learned that the would-be bandit, calling himself "George Anderson," had checked into the San Augustin Hotel more than a week prior to the holdup attempt. "Anderson" disappeared after a couple of days, and the hotel manager entered his room and confiscated a traveling bag. Papers inside the luggage indicated that he was an advance man for the Independent Carnival Company and that his name was "Walter F. Stanley." The hotel manager recalled that the man had complained about being bothered by people to whom he owed money.

On Monday night, word came that a hobo on an east-bound freight train had told other transients in his boxcar that he had a wounded partner in Tucson. Remembering the fourth shot in the fray, Wheeler promptly took a train to the east. On Tuesday night, however, Bostwick died, and orders were telegraphed to Wheeler to return for an inquest the next day. Wheeler reluctantly abandoned his pursuit of the alleged lookout man and headed back to Tucson.[32]

By the time he arrived, it had been learned that the dead man was from Locust Grove, Georgia, and that his wife was in Denver. Telegrams from the deceased's father revealed that his real name was Joe Bostwick, and the elder Bostwick directed that his son should be buried in Tucson.

When interviewed by a reporter for the Tucson *Citizen,* Wheeler commented: "I am sorry that this happened, but it was either his life or mine, and if I hadn't been just a little quicker on the draw than he was I might be in his position now. Under the circumstances, if I had it to do over again I think I would do exactly the same thing." [33]

Harry Wheeler was not the only Ranger forced to shoot a lawbreaker in the fall of 1904. As previously mentioned, Lt. Johnny Brooks felled Charles Douglas with a .45 slug when the fugitive tried to resist arrest in Bisbee. Earlier in the year, John Robinson was killed by Ranger gunfire while resisting arrest in Naco.[34] But Rangers did not always shoot the men they arrested. On Tuesday, December 6, Rangers Billy Old, Chapo Beaty, and James T. Holmes apprehended Antonio Nuñez, regarded in some quarters as "about the most notorious uncaught desperado in Arizona and Sonora." A longtime rustler, Nuñez had the annoying habit of bragging about how many horses and cattle he had stolen. He also frequently boasted that he could never be taken alive.

Sheriffs throughout Arizona maintained a lookout for Nuñez, but early in December the three Rangers tracked the thief near the territory's southwestern border. They sighted him driving stolen livestock, then swiftly and efficiently closed a trap. Despite his boasts, Nuñez surrendered tamely. He was confined in the county jail in Nogales, but the local newspaper fondly predicted that "Sheriff Turner will place him in Yuma for the New Years." [35]

This arrest helped close 1904 on a successful note for the Rangers. Despite a few sour incidents during the year, steady gains had been made against rustling, and violent Ranger action against Billy Stiles, Burt Alvord, Joe Bostwick, and other badmen had served as lessons that could not be ignored by would-be outlaws or the Arizona public.

# 1905:
# Riding Into Danger

*"Were the Arizona Rangers patriots or fools?"* — Lt. Harry Wheeler

By 1905 the Arizona Rangers had achieved their initial objective of curbing rustling activities in areas that had been plagued by stock theft. A special trouble spot had been the Arizona-New Mexico line, where rustlers had slipped elusively from one jurisdiction to another. The Rangers' diligent efforts had driven most stock thieves from eastern Arizona, which created a problem for New Mexico.

So many rustlers now haunted New Mexico that officials decided to follow the Arizona solution. On January 16, 1905, the opening day of New Mexico's Thirty-sixth Legislative Assembly, Governor Miguel A. Otero addressed the legislators by describing his proposals. He outlined a Ranger law, concluding with the observation: "A law of that kind is reported to be working very satisfactorily in the neighboring Territory of Arizona." Within a short time the New Mexico Mounted Police was voted into existence, "identical in all but the name to . . . the Arizona Rangers." The first captain was John Fullerton, and the force served New Mexico until it was discontinued in 1921.[1]

Meanwhile, back in Arizona, it was business as usual for the Rangers. A complaint arrived at headquarters from Fort Apache that Indians on the reservation were losing horses to rustlers. In mid-February, Captain Rynning ordered Jeff Kidder, Joe Pearce, and Oscar Rountree to investigate, admonishing them not to go "up there to build your reputation as gunmen." Pearce felt that this warning was aimed primarily at Kidder,

77

*Jeff Kidder (third from right, astride mule) "& other Rangers (not identified)."*
*Other photos, however, indicate the following identities: Capt. Tom Rynning is at far*
*right; Bob Anderson is second from left; Chapo Beaty, third from left; W. W. Webb,*
*sixth from left.*

"who was a gunman as well as a law man and in a pinch would shoot first
and ask the questions later on, if there was anybody left to answer
them." [2]

The three Rangers put together a pack outfit, consisting of bedrolls,
coffee, jerky, bacon, flour, soda, and butter, then donned heavy woolen
jumpers beneath their coats. They rode north across the Sulphur Spring
Valley to Willcox, pushed on through Solomonville to Clifton, then
headed up the San Francisco River and north along the Blue, riding day
and night with only a few hours of sleep at a time.

They were in mountains now, wild country only thirty miles or so
northeast of where Carlos Tafolla had been killed in 1901. Snow flurries
swirled around men and horses, and the wind in the high country was
hard and biting. The Rangers rode 200 miles in three days, and now they
began to stop in at ranches near the head of the Blue River. Inquiries revealed that a number of ranchers had lost saddle horses.

Toles Cosper, owner of a large spread along the Blue, invited the
Rangers to stop in for supper and a night's stay in the bunkhouse. Two
days earlier, one of Cosper's cowboys had spotted three riders driving
what looked like Indian horses toward the east. One of the men was de-

scribed as tall, slim, and well dressed, while his two companions were short, skinny, flat-headed, long-haired, and thick-lipped. The Rangers put up their spent animals, and in the morning Cosper loaned them his strongest mounts.

The Rangers headed east into rugged terrain, hampered by alternating rain and snow. Near the Arizona-New Mexico border they picked up a trail which led to a campsite in a pasture. The horse droppings were only a day or so old, and three-quarters of a freshly butchered yearling steer had been abandoned by a man obviously in a hurry.

Kidder, Pearce, and Rountree pushed on into New Mexico's San Francisco Mountains. A twenty-mile ride brought them close to the San Francisco River, where the lawmen sighted a lone rider off to the east. Pearce zeroed in on him with his old-fashioned spyglass and made out a tall, slim, well-dressed man. Kidder and Rountree each took a look through the glass.

"Let's get him," said Kidder, spurring his mount. But the rustler spotted the lawmen as they started toward him, and he galloped away toward Arizona. The Rangers kept the man in sight for four or five miles, then lost him in thick timber. When they picked up his trail, they found that he had doubled back to his companions with the stolen horses. There were about two dozen unshod Indian ponies, along with eight ranch horses.

To the Rangers' surprise, the rustlers again turned back toward Arizona. At noon the officers discovered a little valley where the rustlers had taken time to build a fire to cook lunch. Within a couple of hours the Rangers caught sight of their prey, just 300 yards away. Rountree impulsively shouted, "You better surrender, you bastards."

Pearce brought up his spyglass again, but as he did the rustlers, who had no rifles, cut loose with their revolvers. The range was too great, but spent slugs from linear shots dropped nearby. "Down!" warned Rountree, and the Rangers pulled out their rifles and dismounted. But the rustlers already were moving out, driving the stolen horses into the mountain wilderness of eastern Arizona.

The Rangers had no trouble keeping up with the men encumbered with a horse herd. Two or three times they drew close enough to glimpse the outlaws through the rugged countryside. As they neared the Blue River, the rustlers abandoned their stolen animals, thinking that the men on their trail were ranchers who just wanted their horses back. The Rangers passed the grazing horses and pursued the fleeing rustlers. A steady rain soaked their clothing, and the cold wind chilled them to the bone. At dusk the shivering officers lost the trail and sought refuge in a deserted cabin. They hobbled their horses, built a roaring fire, and dried out their clothing. As rain dripped through the dilapidated roof, the Rangers ate and smoked and discussed the next day's chase.

As the eastern sky began to lighten, the three officers quartered for the remains of the washed-out trail. Riding at northeastern and southwestern angles, the Rangers soon discovered tracks that continued west. At sunrise they sighted a log cabin in a valley just three or four miles from the Ranger campsite. Smoke curled upward from the chimney.

The Rangers tied their horses to pine trees, unlimbered their rifles, and crawled through bushes and high grass toward the cabin. After 400 yards on their bellies, the lawmen stopped behind a tree stump to wait for the rustlers to emerge from the cabin door. The Rangers built cigarettes and patiently watched the door.

Soon they could hear voices inside. Then the tall rustler and one of the short outlaws walked outside, holding steaming cups of coffee while they studied the skies. Both men wore six-guns.

Suddenly, the Rangers stood up and leveled their rifles. "Throw up or we'll cut you in two," shouted Rountree.

The shorter man dropped his coffee cup and reached for the sky. The tall rustler stood calmly and asked, "Who are you?"

"We're Arizona Rangers," Rountree answered. "Throw up and don't ask so many questions."

The second rustler raised his hands, and Rountree ordered him to tell the third outlaw to come outside. When the man walked out, the Rangers closed in. "For God's sake," exclaimed the tall rustler, "don't shoot. We won't make no trouble for you."

"You bet you won't," growled Kidder. "You made enough of it already." [3]

Kidder and Pearce kept their rifles on the rustlers while Rountree approached them from behind. He disarmed them the way he had been trained by Rynning, and the guns and cartridge belts were stacked in a pile. Rountree ordered them to remove their hats, boots, and shirts, but a thorough search produced merely a pocketknife.

The Rangers took their prisoners to a justice of the peace named Bockley, whose office was on the Blue River. After a preliminary hearing, it was decided that Kidder and Rountree would transport them to Clifton and turn them over to officers there. Pearce and a couple of ranchers backtracked and rounded up the stolen horses, still grazing in the valley where they had been abandoned. The eight shod animals were returned to the three spreads from which they had been stolen, while a cowboy was assigned to drive the Indian ponies back to the reservation.

By February 1905 Governor Brodie had decided to accept an appointment as assistant chief of the Records and Pensions Bureau of the War Department. President Roosevelt's offer was too tempting to refuse. On February 14 Brodie left office after nearly three years as governor. Although he was a Republican, Brodie was so respected that the Democratic legislature presented him a saber.

President Roosevelt selected Arizona's attorney general, Joseph H. Kibbey, to fill the governor's office. Kibbey had an active career as a lawyer and politician in Arizona and was regarded even by political oppo-

*Officers of the federal court and grand jury at Florence gathered in front of the court-house in 1891. Standing fourth from the left is Joseph H. Kibbey, judge of the district court and future governor of the territory. U.S. Marshal Bob Paul is third from left.*

nents as "a man of good ability and of good reports." [4] He assumed the duties of his new office on March 7, 1905.

The Twenty-third Legislative Assembly already was in session, and among the legislation they considered were three separate bills to abolish the Rangers. In every legislative session after the Rangers were created, at least one bill to abolish the force was introduced. The cost of maintaining the Rangers was running close to $3,000 per month, and by 1905 not a single member remained in either house from the 1901 body that authorized creation of the force. Although some newspapers strongly supported the Rangers, others editorialized against the company — some clamoring for abolishment and some suggesting that the company be reduced by half.[5]

But Ranger arrest totals were mounting at a record rate: during the fiscal year which ended June 30, 1905, the Rangers reported 1,052 arrests.[6] Eight Rangers regularly worked with the Livestock Sanitary Board and the Arizona Cattle Growers' Association, and the force had begun to concentrate on assisting federal authorities in halting the flood of Chinese aliens being smuggled into Arizona. Rangers were arresting swindlers, forgers, army deserters, drunks, bunco artists, vagrants, and men who ran opium dens.

In his first annual report, Governor Kibbey effusively praised the

Rangers: "Life and property are safe in all parts of the Territory . . .";
"Probably the greatest benefit to the Territory from the ranger force is the
fear implanted in criminals . . ."; "criminals . . . know that the rangers, in
pursuit of men charged with crime, are relentless and persistent." Kibbey
realized that a major source of opposition to the company was money:
"While expensive, the ranger force has accomplished excellent results."
The governor pointed out that the spread of railroads throughout the territory would make crime easier to control, "and I am hopeful that within
the near future the Territory can be warranted in reducing the force and
thereby lessen materially the cost of its maintenance." [7] The support of
the new governor and, most importantly, the impressive performance by
Rynning and company, caused all efforts to abolish the Rangers in 1905 to
fail.

In the spring of 1905 some Rangers executed an arrest which doubtless won them little affection from many of their fellow officers in Arizona.
Lee Hobbs, who had been the chief deputy sheriff of Graham County for
a year, was stationed in Clifton, and his brother, also a deputy sheriff, was
stationed at nearby Morenci. The Hobbs brothers were nephews of Sheriff
Jim Parks, and although Lee had shot a man to death shortly after his appointment, he retained his commission and never even went to trial. Lee
had "the reputation of being a bad man," commented a newspaper. Sgt.
Billy Sparks, a Ranger stationed in Globe, apparently felt no reluctance
when presented the opportunity to end Hobbs's career in law enforcement.[8]

"Timberline Bill" Sparks worked up a case against Hobbs for mutiny
and murder on the high seas. In 1902 Hobbs had signed on to work
aboard the British freighter *Lancaster Castle,* which was steaming from San
Francisco to the Orient. En route Hobbs, becoming disgruntled, led an
uprising and shot the captain and first mate, killing the latter. Hobbs and
his chief accomplice escaped in a small boat to a Pacific island, then took
passage on the first steamer which happened by. Hobbs disappeared for
two years, even though British officials posted a large reward and initi-
ated an exhaustive search.

Sergeant Sparks's primary clue to the involvement of Hobbs in the
mutiny came from a chance meeting on an Arizona train with a man
named Walz. Walz, the cook on the *Lancaster Castle,* went to Clifton and
gave Sparks a positive identification of Hobbs, who had turned up in Arizona in 1904. English officials in San Francisco were contacted, and in
April Consul General Courtney Bennett and T. K. Carmac, an English
attorney practicing in San Francisco, traveled to Phoenix and secured extradition papers. Sparks met Bennett and Carmac in San Francisco, then
the three men journeyed together to Clifton.

Realizing the explosive potential of arresting a fellow officer who had
many influential relatives in the area, Sparks requested help. Soon Sgt.
Harry Wheeler and Privates Jeff Kidder and Oscar Rountree gathered in
Clifton. When the four well-armed Rangers confronted Hobbs on Friday

night, April 7, there was no trouble, but problems were expected. Fears of a violent rescue proved groundless, at least in part because of the formidable Ranger presence, and Hobbs was extradited for trial.

But hard feelings against the Rangers increased. Private Rountree, while making an arrest in Graham County, was charged with assault. After securing Rountree's conviction, the district attorney "stated that the rangers are not wanted in Graham county, and that they have no business there, as the officers of the county are able to take care of the criminals of that county, and that the Graham county citizens are very much against the force." [9]

Various other Rangers also handled noteworthy cases during the year. Sgt. Billy Old and Chapo Beaty captured notorious horse thief Simon Juarez, who soon was on his way to a cell at Yuma.[10] Later in the year Sergeant Old teamed with Pvt. Garland Coffee to capture Luis Rivera, a murderer, horse thief, and jailbreaker.

In 1903 Rivera was convicted of murder for hire in Sonora and sentenced to fifteen years in prison in Altar, sixty-five miles southwest of Nogales. Rivera escaped in 1904 and for a year and a half slipped back and forth across the border, eluding officers from Sonora and Arizona and bragging that he would never be taken alive. But on September 26, 1905, Old and Coffee sighted him in a horse corral as they rode up to a ranch in the Oro Blanco Mountains, west of Nogales. Old instantly spurred his mount, galloping forward to block the corral gate. Old's Winchester was out, and he shouted at Rivera to throw up his hands. A six-gun belted to his waist, Rivera ignored Old's command, smiled defiantly, and sneered, "Go on and shoot."

"All right, old man," snapped Old, throwing his Winchester to his shoulder and sighting in between Rivera's eyes. "I'll shoot you quick enough if you don't get your hands up."

Rivera saw death in Old's eyes. He threw his hands upward, trembling visibly. When Coffee cuffed his hands they were still shaking. Rivera was taken to jail in Nogales prior to his extradition — an event no doubt anticipated eagerly by his jailer in Altar. After Rivera's escape, the jailer had been convicted of negligence and sentenced to serve out Rivera's term unless the escaped convict should be returned to prison.[11]

Other cooperative ventures along the border in 1905 included a crackdown on munitions sales to the murderous Yaqui Indians. Yaquis were on a rampage in Sonora and were purchasing guns and ammunition from merchants in Nogales, Bisbee, Douglas, and other border towns. Following complaints from Mexican authorities to Arizona officials, in June the Rangers "received strict instructions" to stop the arms trade to the Yaquis.[12]

Also in June, Lt. Johnny Brooks rode into Sonora after rustlers who had stolen seven horses and mules from Pirtleville, a few miles west of Douglas, then headed for nearby Mexico. Accompanying the Ranger lieutenant were his brother Jap Brooks and Deputy Sheriff Young Davis. The trail led southwest, and the little posse rode hard, covering hundreds of miles of desert and mountain terrain in three days. The Americans caught up with their prey below Magdalena, capturing one rustler and recovering several head of stolen stock. They herded the animals north, leaving the recovered stock and their spent mounts in Nogales while they returned by train to Douglas. Lieutenant Brooks bound over the rustler "to the tender mercies of the Mexican authorities, and as he had a bad record, he has probably passed away to the happy hunting grounds where all good horse and cattle thieves go, when they are lined up in front of a bunch of Rurales." [13]

In October Captain Rynning was granted permission to go into Mexico to bring back the body of S. L. Riley, a mining engineer who had been murdered at Simola. After Riley's murder he had been hastily buried, and Mexican law mandated that a corpse could not be taken out of the country until it had been interred for five years. However, Riley was heavily insured and the insurance could not be claimed until the body was produced. As another example of Sonoran-Arizonan cooperation, Mexican authorities agreed to waive the law if Rangers would claim the corpse, and Rynning sent Harry Wheeler and Arthur Hopkins "into the wildest region of Souther[n] Sonora" to retrieve Riley's remains. [14]

Another October incident involved the bloodiest shooting in the recent onslaught of violence at Silverbell, a mining camp northwest of Tucson. On Monday night, October 2, four Anglos left Webster's Saloon, and as they walked down the street they passed three Mexicans. One of the Mexicans, apparently drunk, was being supported by his two *compañeros*. Suddenly, without evident provocation, the Mexicans opened fire at the backs of the Anglos. Three of the men were wounded, two fatally, while the murderers fled on foot into the night.

Ranger Reuben Burnett, recently stationed in the raucous mining camp, was summoned, and he hurried off in the direction of their escape. It was too dark to follow a trail, however, and Burnett's search proved fruitless. The next day Sheriff Nabor Pacheco arrived from Tucson, while the two corpses were shipped by train to Reilly's Undertaking Parlor in Tucson. Pacheco and Burnett headed up a massive search effort, while in Nogales Rangers Billy Old and Garland Coffee headed west to intercept the killers in case they tried to go to Mexico. The countryside was combed for three days, but on October 6 Pacheco and other officers found the men accused of the crime. Meregildo Rodriguez and Manuel Parra had hidden in Silverbell and were at the depot hoping to leave by train when they were apprehended. [15]

There was talk in the summer of 1905 that a change in Ranger leadership could take place. Captain Rynning apparently was uneasy for a time after his Rough Rider companion-at-arms, Brodie, left the governor's office for Washington. Soon there were rumors that Rynning would resign from the force, and that "Rangers Wheeler, Brooks, Sparks and Holmes are candidates for the captaincy." [16] In Bisbee it was predicted that Lieutenant Brooks "will succeed Captain Rynning when the latter goes to Yuma as superintendent of the Territorial prison." [17]

Rynning evidently had been told by Governor Brodie that he would be appointed superintendent of the territorial prison at Yuma when the position next became vacant. After Brodie left office, Rynning assumed that the new governor likewise intended to elevate him to the the superintendency. When Superintendent Ben F. Daniels resigned, the Ranger captain did not even bother to make a formal application. Governor Kibbey, who had never been told that Rynning wanted the job, appointed Jerry Millay to head up the prison. [18]

Rynning shrugged off his disappointment without public complaint, and soon he was preoccupied by a clash with his second-in-command. Johnny Brooks, who was regarded by Rynning as "brave, loyal, and . . . honest as a horse," was offered a cash bounty to lead a movement against a mine across the border in Sonora. There was a dispute about the ownership of El Tigre Mining Company, and a group of Kansas City stockholders engaged the capable Brooks to seize the mine on their behalf. When Rynning discovered that his lieutenant was about to lead a foray into Mexico, a "friction" erupted between the Ranger officers. Brooks resigned on July 6, 1905, then became a special officer for El Tigre Mining Company. [19]

By this time Sgt. Harry Wheeler had impressed everyone with his efficiency and almost fanatical devotion to duty, and Captain Rynning quickly appointed the diminutive Ranger to fill the lieutenant's post. Billy Old, whose leadership qualities also were obvious, was promoted to sergeant to replace Wheeler. [20]

Lieutenant Wheeler was stationed in Willcox, but he had been on the job only a few days when he accompanied Rynning by train to Yuma on official business. Shortly after his return, the new lieutenant was dispatched by rail to Prescott to help Pvt. Oscar Rountree smooth over a local difficulty. [21] Hardly had he again returned before Lieutenant Wheeler learned that a more demanding assignment awaited him in the weeks ahead. He would be filling in for Captain Rynning, who planned a marathon trek to inspect Ranger outposts across Arizona "and all the large cities of the Territory." [22]

In mid-August Rynning strapped his bedroll and ample provisions onto a packhorse, saddled his big gray, then mounted up and headed west for a journey of perhaps 1,500 miles. Although Rynning kept in touch with headquarters, Harry Wheeler ran the company for the next six weeks, experience which would prove invaluable within a couple of years.

Captain Rynning rode to Bisbee, checked out the Ranger station in Naco, then moved into Santa Cruz and Pima counties. He continued in a westerly direction, meeting with Rangers at each station before riding back to Douglas.

He had traveled 500 miles when he procured fresh animals and ventured toward the north. The Rangers had received complaints from ranchers in northeastern Arizona that cattle were being killed on the range, apparently by disgruntled Indians. Rynning intended to investigate the situation personally while continuing his inspection of Rangers scattered across the territory.

The captain rode through Cochise and Graham counties, then turned westward into Pinal County. He went north through Gila County, angled in a northeasterly direction across Navajo County, and finally entered Apache County, a long political district measuring approximately fifty miles east to west and 225 miles down (from the northeastern corner of Arizona to the western border of New Mexico). Rynning rambled south through this beautiful but rugged country that had seen so much Ranger activity against rustlers. In the White Mountains he encountered a group of Bisbee men who were on a hunting vacation. Rynning took time out from his official journey to serve as guide for the hunting party.

When he finally made his way back to Douglas, "browned by exposure to the wind and sun, " Rynning had ridden horseback a total of 1,300 miles. He now had a first-hand impression of conditions across much of the territory, and he hastened to Phoenix by rail to report to Governor Kibbey. A man with keen political instincts, Rynning still had his eye on the prison superintendency or some other possible promotion, and he began to communicate cordially with Arizona's new governor.[23]

While in Phoenix, Rynning was interviewed by an *Enterprise* reporter about his six-week inspection patrol. Rynning was modest about his journey, but the *Enterprise,* like many Arizona newspapers, reacted favorably to the popular, efficient Ranger captain. Soon it was business as usual for Rynning, who went on from Phoenix to Prescott on official duties.[24]

During 1905 Rynning enlisted fourteen recruits, but none of the men carved out notable careers with the company. Five of the 1905 enlistees left the force within a few months. Texan Boyd Doak, for example, signed up on August 23, but his personnel report deemed his character as "unreliable" and on September 25 he was discharged for "Drunkenness and Disobedience." [25] Oscar Felton, who had enlisted early in 1902, was discharged in July for disobedience of orders. Nine other veterans resigned in 1905, including Joe Pearce and Bud Bassett, as well as Lieutenant Brooks. Henry Gray, the last of the original Rangers, had served five years, but at fifty-one, when his third reenlistment expired on October 12, he chose to retire from the force.[26]

One 1905 recruit, Henry McPhaul, had a busy fall on behalf of the Rangers. A thirty-eight-year-old Texan from Waco, McPhaul already had considerable experience as a peace officer when he enlisted in the Rangers on August 1 in Yuma. McPhaul had served as a deputy sheriff in Maricopa County, lost the sheriff's election in 1894 by one vote, worked as a prison guard at Yuma until he allowed a prisoner to get past him over a wall, but then wore three different badges in Yuma until his appointment as a Ranger.[27]

McPhaul was stationed in Yuma, where he acquired a reputation for dealing harshly with Mexicans. On Monday night, September 11, he encountered Donaciano Ortego, who was quite drunk and seated on a step in front of Joe Balsz's meat market on Main Street. Ortego, known as "Guero," stood up and cursed McPhaul. When the Ranger marched over to arrest Ortego, Balsz tried to intervene, but McPhaul shrugged him off and grabbed Guero by the shoulder.

"I have to arrest you," announced McPhaul, "come along."

Guero, however, wrapped both arms around an awning post and fired a long stream of profanity at the Ranger. McPhaul, a small man, tugged at Guero in a vain effort to dislodge his tormentor. Then the Ranger pulled his six-gun and applied the barrel several times to Ortego's skull, "leaving Guero's head in a condition decidedly unbeautiful to look upon." Finally persuaded, Guero released the post and was led off to jail.[28]

The next morning Guero pled guilty to disturbing the peace and paid a seven-dollar fine. Then he filed a suit against McPhaul for assault. Disdaining an examination before a justice of the peace, McPhaul posted a $400 bond for a grand jury hearing.

When McPhaul's hearing was held on October 7, before the Honorable J. H. Campbell of Arizona's First Judicial District, the charge was dismissed due to lack of evidence. But the grand jury foreman, a former Yuma County law officer named O. F. Townsend, stood and read a statement by the group: "It is the unanimous opinion of this grand jury that in many cases of arrest, peace officers use undue force where by the use of judgement and discretion it would be clearly possible to conduct offenders to jail without recourse to violence and believes that this matter should be brought clearly and emphatically to the attention of the officers charged with the keeping of the public peace." [29] As long as the force existed, such complaints continued.

Little more than a week after McPhaul was subjected to the grand jury's reproval, the Ranger recognized a thief named Harry Wilson on the streets of Yuma. Wilson had a string of aliases and a notorious reputation, and after McPhaul spotted him he made an arrest. McPhaul turned Wilson over to Constable Julio Martinez, who threw him into jail for vagrancy. As the local newspaper put it, "the Ranger figured that a stitch in time would save nine." Within two weeks McPhaul was transferred to Silverbell in Pima County. Although Silverbell was plagued with shootings, knifings, and miscellaneous violence, McPhaul liked his new station so well that he contemplated selling his Yuma property for a permanent move to town.[30]

James T. "Shorty" Holmes fought the first of three shootouts while wearing a territorial badge in 1905. Holmes had been promoted to sergeant, but by 1905 he had reverted to private, apparently with no hard feelings, perhaps by his own choice.

Holmes was stationed at Roosevelt, east of Phoenix where the massive Roosevelt Dam was under construction. On Tuesday afternoon, October 31, Holmes intercepted Bernardo Arviso, a bootlegger suspected of selling liquor to Indians. Arviso tried to fight his way past Holmes, and a sharp pistol duel erupted. A government teamster named Bagley tried to aid Holmes but caught a bullet from the bootlegger in the arm. The Ranger fired back with deadly aim, killing Arviso on the spot.[31]

Within four months Holmes again engaged in a fatal gunfight near Roosevelt. On February 18, 1906, he clashed with an Apache outlaw known as "Matze Ta 55." The Apache was shot to death, but Holmes won complete exoneration by a coroner's jury. In 1907 Holmes was in action again, this time trading shots with smugglers. During his years as a Ranger, Holmes never suffered a wound, and he was cited for distinguished service in the 1906 and 1907 engagements.[32]

In April 1907, however, Holmes was criticized in some quarters for rough treatment of a prisoner. A black man named Baldwin had murdered a Mrs. Morris and her daughter near Roosevelt on January 31, 1907. Holmes finally apprehended Baldwin just outside Roosevelt. Baldwin wisely surrendered to the lethal Ranger, but Holmes — never kindly disposed toward murderers — beat him over the head with a frying pan. Then he tied a rope around Baldwin's neck, mounted his horse, and spurred away, dragging the prisoner "for a considerable distance." Holmes stopped long enough to tie Baldwin's hands, then dragged him on into town, where the postmaster prevailed upon him "to cease abusing the helpless man." There was a certain amount of newspaper criticism of the Ranger, but Baldwin probably was fortunate he had not been shot by Holmes.[33]

Ranger services were not confined to curbing violence. During the summer of 1905, an expedition to a remote Mexican island resulted in a tragedy which eventually involved Captain Rynning and some of his men. Thomas Grindell, superintendent of the Douglas schools, organized a small band of adventurous men who intended to explore Tiburon Island in the Gulf of California. The volcanic island was inhabited by Seri Indians, who were quite primitive and rumored to be cannibalistic. Grindell had visited Tiburon the year before, and the educator hoped to explore the island and record scientific data. It was also rumored that Grindell was searching for a gold mine.[34]

Grindell and three companions — Olin Ralls, Dave Ingraham, and J. S. Hoffman — entered Sonora shortly after school ended. Col. Emilio Kosterlitzky pleaded with them not to follow their intended route, which included a lengthy stretch of barren desert. His advice was ignored. By the end of August neither Grindell nor any of his friends had been heard from, and his brother, Ed Grindell, futilely conducted a relief expedition. On October 11 Captain Rynning wrote Colonel Kosterlitzky, who replied that he was "positive that the poor fellows died for lack of water." [35]

A few days later J. S. Hoffman emerged from a five-month ordeal in the desert. The survivor related a horrible tale of thirst and desperate wandering about in the desert. At one point Hoffman resorted to drinking

*J. T. Holmes twice killed outlaws in gunfights near his station in Roosevelt.*
(Courtesy Arizona State Archives)

seven cups of sea water. Their Papago Indian guide deserted the party, and the four men became separated. After incredible hardships Hoffman — naked, starving, a mass of running sores, bloated from scurvy, and blackened from exposure to the sun — crawled into Guaymas.[36]

Tom Rynning applied to Governor Kibbey for a thirty-day leave of absence to search for the other three members of the party. There was widespread interest in Hoffman's grisly tale, and editorials already had called for assistance by the Rangers. Kibbey gave permission to Rynning, Billy Old, Tip Stanford, and Reuben Burnett to venture into Mexico as members of a private search party. Rynning and his men did not go in their official capacity as Rangers; a private subscription collected in Douglas provided them with traveling funds.[37]

They reached Guaymas, where they joined Dr. Frank Toussaint and J. S. Hoffman, who had recovered remarkably. Dr. Toussaint of the Guaymas area was a close friend of the Grindells, and he gave Rynning $100 to charter a boat and buy provisions. The *Lolita,* a gasoline-powered boat of twenty tons, was engaged, and late on Thursday, November 2, the search party left for Tiburon Island. Rynning, Old, and Stanford were aboard, along with Hoffman, Dr. Toussaint, and the *Lolita's* captain and five crewmen. Two hours after the *Lolita* left Guaymas, word arrived from Walter Douglas that Rynning could draw $500; later the Ranger captain regretted that he had not received the message, since he would have stayed longer and searched the island's interior for bodies.

The *Lolita* sailed 125 miles to Tiburon Island, arriving in twenty-four hours. The next day, Saturday, November 4, the search party sailed around the island but saw only five deserted Seri villages. The captain and his sailors, obviously terrified of the Indians, flatly refused to set foot on the island. They sailed across to the mainland, where they went ashore and were guided by Hoffman to the Grindell party's last campsite. They found two dead horses and a dead burro, saddles, camp equipment, miscellaneous provisions, 300 rounds of ammunition, and a camera belonging to Grindell. They scouted trails for several miles but found no trace of the missing men or their bodies. Concluding that there was nothing further to accomplish near the campsite, Rynning and his men returned to the *Lolita* and headed back to Guaymas.[38]

Later Ed Grindell led another scout party, but by this time heavy rains had obliterated all trails and the search proved fruitless.[39] Eventually everyone agreed that Tom Grindell, Ralls, and Ingraham had died of thirst. Their corpses evidently had been taken by Indians and probably would never be found. "I have given up all hope," said Ed Grindell, "that my brother or any of his party unaccounted for are yet alive." [40] The Rangers had played a significant role in satisfying Arizona that nothing further could be done in solving one of the great tragedies of the day. Billy Old soon would return, along with three other Rangers, on another privately sponsored campaign into Sonora.

Late in 1905 four Rangers performed a mission of heroic proportions, combining detective skills, selfless devotion to duty, and physical stamina in the face of incredible danger. Earlier that summer, a grisly homicide was committed at a small ranch on Pinto Creek one mile outside the village of Livingston. On the dark, rainy night of July 12, rancher Sam Plunkett and an elderly friend, Ed Kennedy, were slain in their beds. Their skulls were crushed with a piece of iron from a wagon tongue, and Plunkett was stabbed sixteen times. Kennedy also was stabbed repeatedly. They lay dead for two days before their corpses were discovered. Robbery was apparent; about $100 in cash was missing, as well as a gold watch and a revolver. Two Mexican employees — a big man named Gonzalez and a youth named Ascension — were missing and presumed guilty of the crime.[41]

Pvt. William S. Peterson, a Texan who had been in the Rangers since 1902, headed the immediate investigation. Aided by Indian trailers and Al Sieber, the noted army scout of Indian war fame, Peterson followed the tracks of the suspects for twenty miles in the direction of Globe. The two Indians circled in opposite directions until they cut trail. The killers had been on foot, but ten miles from Globe they apparently were picked up and rode away on horseback. Although the trail disappeared in Globe, a blood-stained knife was found, along with a bloody shirt which had been discarded.

Soon it was learned that Pantaleon Ortega, who tended bar at Mark Cheever's boardinghouse in Globe, had helped the two criminals escape. Evidently he brought his two friends to the boardinghouse, tended to three knife cuts suffered by one of the men in his leg, cashed a paycheck for them, and bought them new clothing. Ortega secluded Gonzalez and Ascension for two or three days before helping them to leave the vicinity.

Law officers, apparently including Lt. Harry Wheeler, noted that whenever a man with a badge was present, Ortega was noticeably uneasy. Mrs. Cheever was asked why the bartender was so nervous around lawmen, and she went to Ortega. He confessed to her about assisting his two friends, whom she remembered from several meals at her boarding table. Ortega also told her that he had received letters from them from Minas Prietas in Sonora. Mrs. Cheever relayed the conversation and stated that she was certain she could recognize the men. Harry Wheeler had Ortega arrested and placed in the Gila County jail in Globe. It was learned there that one of the suspects had been confined overnight for vagrancy after the murder, but he was released routinely the next morning.

Months passed with no further progress in the case. However, even as the crime faded from public attention, Lieutenant Wheeler and a few other law officers never stopped probing for a lead. At last Wheeler heard from an informant that the two suspects were in Sonora, going from one mining camp to another and never staying more than two weeks in one place. Wheeler, Sgt. Billy Old, and new Ranger recruits Dick Hickey and Eugene Shute determined to go into Sonora in search of the killers. Shute was related to one of the victims, Plunkett, and he may have joined the Rangers for the sole purpose of seeking his kinsman's killers in Mexico. Hickey, too, may have enlisted primarily to go on this manhunt, perhaps in hopes of reward money, which by this time totaled $1,050.[42]

In Magdalena there was no sign of the fugitives, but Wheeler and his men paid a visit to Kosterlitzky, asking the *Rurale* commander for aid in finding the murderers. The *Rurales* had no men to spare, however; the Yaqui Indians were engaged in a typically ferocious uprising, and Kosterlitzky warned the Rangers to be wary of these murderous tribesmen during their search.

Wheeler, Old, Hickey, and Shute took a train to nearby Santa Ana, then traveled by stagecoach to a remote mining camp. The fugitives were not there, so the Rangers returned to Santa Ana and boarded a train for a 150-mile trip south to Ortiz. In Ortiz the Rangers bought supplies and acquired a wagon and team, intending to proceed to La Dura, ninety miles distant and located in the heart of Yaqui country. But as they were preparing to leave the next morning, a Mexican army officer arrived and forbade their journey. He stated that a Yaqui war party was within ten miles of Ortiz and the roads were unsafe. A cavalry squadron was camped nearby, however, and he promised to provide them with a military escort.

The day passed with no sign of an escort. Late that night the Rangers quietly hitched up their horses and headed toward La Dura, where they

*Harry Wheeler, newly appointed lieutenant of the Rangers, led an epic 1905 search into Mexico for murderers.*                    (Courtesy Arizona State Archives)

hoped to find Gonzalez and Ascension. After driving about twenty miles they encountered a battered cavalry troop whose commander described a vicious fight the day before, with severe losses to the Yaquis. He detailed the dangers of the country ahead and tried to dissuade the Americans from proceeding. Wheeler, however, called him aside, and with his characteristic sense of obligation explained that he and his little squad also were soldiers, as loath to turn back from their duty as soldiers in Mexico. The Mexican leader gravely saluted Wheeler, and the two bands of men went their separate ways.

After making good progress, the Rangers camped for the night in a cluster of mesquite trees. In late December on the western approach to the Sierra Madres, nighttime temperatures plunged to chilling levels, but the danger of discovery by Yaqui warriors made a campfire too great a risk. The Rangers, who had no blankets, spent a miserable night shivering and stamping and waiting for the dawn so that they could again be on the move.

At daybreak, however, another cavalry troop rode up and took them into custody: they would be guests of the military for their own protection. Wheeler protested, explaining why they were in Mexico and that they would assume all responsibility for their actions. Nevertheless, they were escorted to a blockhouse five miles away, where a small garrison provided them with beds and food.

After four days at the blockhouse, letters from Ortiz ordered the release of the Rangers and the protection of an escort. Wheeler and his men traveled to La Dura, where a search for the Plunkett-Kennedy murderers proved fruitless. They returned with their escort to the blockhouse after a round trip of ten days, then proceeded east a short distance to Guaymas. The fugitives were not in Guaymas, however, nor in Empalme, a few miles to the west.

Doggedly the Rangers decided that the murderers might be working with crews constructing a railroad across one of the most deserted stretches of Sonora. They rode a train twenty miles to the end of the completed track, after which they set out on foot. The Rangers trudged far into the night, then spent a few miserable hours trying to sleep through the cold. The next morning they ate the last of their provisions, put themselves on short water rations, and struck out along the right-of-way. During the day's trek Hickey became so footsore that he removed his shoes, tied the laces together, and slung them around his neck. His three companions laughed at him, but the next day they each had to resort to the same technique.

By this time their canteens were empty, and thirst was added to the misery of hunger and fatigue. On their third day afoot the four Rangers stumbled ahead without speaking, carrying empty canteens and stopping to rest every ten minutes.

For three days they had encountered no one and seen not a single habitation. Wheeler, realizing how desperate their situation was, ordered his men to push on through the night. They staggered wearily onward in silence that finally, gloriously, was broken by an excited cry from Hickey: "That's a house!"

A hint of light flickered through the darkness. The Rangers plunged toward the light, soon coming to a one-room adobe shanty. The only openings were a door and a few portholes. When the Rangers knocked on the door, feet scurried inside and a high-pitched voice dubiously proclaimed in Spanish: "Get your pistols, boys. Kill the Yaquis."

Shouted explanations from the Rangers only produced such a cacophony of screams, curses, and prayers from inside that the desperate Americans could not make themselves understood. Braving the possibility of gunfire, the Americans burst into the now darkened room and struck a match.

Seven men, three women, and half a dozen children cowered in a corner. The men, although terror-stricken and without a single gun, had knelt in front of the women and children, intending to be the first to be slain by the dreaded Yaquis.

Wheeler and the others tried to explain that they were not Yaquis but found it almost impossible to penetrate the panic of the Mexicans. At last, however, the Mexicans realized that the intruders meant them no harm, and the Rangers made clear their identity and plight. The relieved Mexicans became perfect hosts, placing the Rangers on their own pallets, then carefully allotting them water in gradual quantities. The *norteamericanos* had their battered feet washed and bandaged, after which the women served tortillas, beans, goat's milk cheese, and coffee. The exhausted Rangers then fell into a deep slumber, not awakening until the middle of the next afternoon.

The Mexicans continued to tend the Rangers, explaining that the main road crew was ten miles farther on but would return within a few days. The Anglos availed themselves of their hosts' hospitality for three days, as their feet and bodies recovered from the recent ordeal. When the construction crew came back, Gonzalez and Ascension were not among them, so the Rangers made their way back to Empalme.

Next they decided to try their luck at Minas Prietas, where a solid clue finally emerged. As the Rangers searched the camp, they were told that two men fitting the description of Gonzalez and Ascension had left Minas Prietas just a few days earlier. They had announced their destination as Nogales, and the Rangers immediately headed north.

The suspects could not be located at Nogales, but the entire Ranger force was alerted that the murderers probably were back in Arizona. Mining camps throughout the territory were scoured in vain. Wheeler returned to duties in Willcox — and happened upon the very men he had trailed back and forth across Sonora.

It was raining hard in Willcox, but despite the deluge two Mexicans sat on the railroad tracks outside town, refusing to come in for shelter. Wheeler's suspicions were aroused, and he went out to confront the two drenched men. When Wheeler approached, the two men stood up and began to move away. The Ranger lieutenant called out that they were under arrest. The larger of the two men, Gonzalez, whipped out a wicked-looking knife, and his *compañero*, Ascension, opened up a pocketknife. Wheeler pulled a gun and easily took them into custody.

Gonzalez and Ascension were jailed in Willcox, and they nervously paced in their cells throughout the night. Gonzalez was overheard repeating, "We will be hung in Globe." But Ascension always replied, "No. They will hang you but not me." Ascension later stated that Gonzalez had slain both Kennedy and Plunkett. Gonzalez had forced Ascension to watch while he stabbed and battered their employers, then the two thieves seized the money and divided their plunder before fleeing on foot. When the two were bound over to the Gila County grand jury, Ascension confessed willingly and wangled a release. Gonzalez managed to escape, but later committed suicide.

Harry Wheeler was understandably disgusted at these developments. He and his three companions not only had braved enormous hazards to search a foreign, hostile land for the murderers, they had received no travel allowance from the territory. Relatives of the dead men provided some of the expense money; otherwise the Rangers paid for their own way.

Recruits Hickey and Shute quit the Rangers shortly after returning from Sonora. Hickey resigned on January 20, 1906, and Shute left the force eleven days later.[43]

Harry Wheeler, reflecting in 1910 on the mammoth manhunt, marveled at the dedication of his men: "Every conceivable hardship and danger had been endured uncomplainingly. Rangers drawing $100 per month, with absolutely no expense allowance, had readily and willingly spent their own wages . . . in the performance of their duty." With a trace of rueful enlightenment, he asked rhetorically: "Were the Arizona Rangers patriots or fools? I leave it to the questioner." [44]

# 1906:

# Maintaining a Stronghold

*"Yes, I guess I have had a ton of lead fired at me one time or another."* — Pvt. Sam Hayhurst

Harry Wheeler found himself unusually busy during the early weeks of 1906. As the new year began, a criminal gang exerted such a grip on Douglas that it became dangerous to go out at night. It was an embarrassing reflection on the Rangers, since headquarters was at Douglas, but Tom Rynning simply turned the problem over to his capable lieutenant.

Wheeler characteristically met the criminal challenge head-on. He brought six Rangers into Douglas, then led them in patrolling the city streets after dark. While catching some sleep during the day, Harry Wheeler and his men rode horseback for ten consecutive nights through the streets of Douglas from dusk until dawn. The "yeggmen" (slang of the day for criminals) were completely cowed by the imposing presence of the heavily armed Rangers astride their horses. The Ranger patrol "put the thieves and thugs to flight," intimidating the yeggmen so thoroughly that Wheeler and his men could find no one to arrest. A newspaper headline trumpeted: "Rangers Rescued Douglas From Band of Yeggmen." [1]

Within days of taming Douglas, Lieutenant Wheeler was in Tucson, where he arrested a fugitive named Harry Howard. Howard, still a teenager, had been incarcerated for burglary in what Los Angeles boasted as the finest jail in the Southwest. But Howard and another inmate found escape easy: they crawled out through a hole in the wall on the second floor and lowered themselves with a rope. Howard's description was widely circulated, and a reward of $100 was offered.

The escapee slipped into Tucson aboard a train on Sunday afternoon, January 28. After dark he wandered from the railroad yards uptown, hoping to find food. At about 8:00 he was at Congress and Church streets, where he was noticed by Harry Wheeler. Wheeler, already searching for the prisoner, thought he recognized Howard and took him to jail. Los Angeles authorities were notified, and Wheeler picked up $100 "Pin Money."[2]

Another 1906 incident involving Wheeler occurred when an American wool grower whose ranch was near Naco discovered that several of his goats had strayed across the line into Mexico. He crossed over into Naco, Sonora, and asked a Mexican customs official named Jiminez for permission to search for his goats. Jiminez granted his assent, but while the rancher hunted his goats, Jiminez turned up and placed him under arrest. The luckless wool grower was tossed into Naco's *juzgado local*.

Lieutenant Wheeler was in Naco, and the rancher had informed the Ranger of his little foray into Mexico. When Wheeler learned of the rancher's arrest, he confronted Jiminez in the Mexican customs office. Jiminez told the Ranger to remove his hat. Wheeler refused unless Jiminez would remove his headwear. There was a tense standoff, then Wheeler stalked out of the office — with his hat still settled squarely on his head.

Wheeler promptly notified Captain Rynning in Douglas about the incident. Rynning contacted Gen. Luis E. Torres in Mexico, who wrote to the Ranger captain. Torres first expressed gratitude for the recent help of the Rangers in watching the border for a gang of *bandidos* reportedly heading toward Arizona. Then Torres stated:

> I deeply regret the lack of courtesy shown by the administrator of the customs service at Naco toward my friend, Lieutenant Wheeler, and I have reported the case to the proper officials. I wish to apologize myself for the lack of good behavior on the part of Mr. Jiminez, and I wish to say further that I work at all times to keep the best of relations between the people of our two countries.[3]

The company continued to take the field against stock thieves. In March, five Rangers and three other law officers operated against suspected rustlers in the southwestern part of Pima County. There were a number of Papago Indian villages in the area. Although many Papagoes were successful horse and cattle raisers, some were suspected of rustling. Sheriff Nabor Pacheco, who worked well with the Rangers, asked Captain Rynning for assistance, and an unannounced sweep through various Papago villages was planned.[4]

Rynning himself rode in the posse, along with Billy Old, Jeff Kidder, James McGee, and Ray Thompson, while Sheriff Pacheco brought two of his deputies. The posse crossed the Baboquivari Mountains and began to ride from one village to another. The unexpected arrival of eight armed riders caused great excitement in each village. Posse members ducked into adobe huts, looking for hides that might give evidence of cattle rustling.

They rode to the border and ranged as far as 100 miles southwest of Tucson, but to no avail. There were no indications that the Indians had stolen any livestock, and after several days in the field the posse turned back toward Tucson. The Rangers had made a concentrated effort against rustling for years, and by 1906 a far-ranging expedition could not produce even one stock thief.

The action which produced the greatest Ranger notoriety of 1906 involved an international incident and the Southwest's most flamboyant promoter. "Colonel" William C. Greene (the title was nonmilitary, assumed by the promoter to enhance his image) had based his fortune and business enterprises on the copper mining district of La Cananea, thirty miles below the Arizona-Sonora border. A larger-than-life individual, Greene was born in Wisconsin in 1853 and went west to seek fortune when he was nineteen. He was a surveyor, teamster, farmer, and cowboy in North Dakota, Montana, Kansas, Colorado, and Texas. By 1877 he had drifted to Arizona, where he prospected relentlessly for mineral wealth, fought Indians on several occasions, and gambled inveterately at poker, faro, and roulette.

Greene married a prosperous widow and invested her inheritance in cattle. When one of his daughters drowned, he shot to death a neighbor he regarded as indirectly responsible. Then he suffered the demise of his wife two years later. Investigating old gold and silver mines in Sonora's Cananea Mountains, Greene discovered quality copper ore that he was able to develop with a series of dazzling promotions. Attracting well-heeled Eastern investors, Greene organized the Cananea Consolidated Copper Company, with himself as president and general manager. He later built a railroad from Ronquillo, where the smelters were located, to the nearest connection in Naco. In 1902 he spent $50,000 erecting a thirty-four-room mansion in Cananea; the elegant residence boasted stables and an Italian sunken garden.[5]

Greene's mansion was located at La Mesa, the American residential district on a mesa northeast of Ronquillo. At the base of the Cananea Mountains, Ronquillo consisted of saloons and dance halls, a big company store, the smelting plant, offices, miscellaneous shops, and the modest houses of Mexican miners. Ronquillo and La Mesa were connected by a railroad bridge across an arroyo. Copper was dug throughout the district at Chivatera, La Democrata, Capote, and several smaller camps scattered across the mountains.

Cananea Consolidated employed thousands of miners and paid them well, by standards of the day. Americans drew union wages in gold and served as shift bosses and in numerous other key positions. But the heart of Greene's labor force consisted of cheap Mexican labor, which permitted him to undersell his copper competitors in the United States. The Mexi-

*"Colonel" William C. Greene, the flamboyant promoter who built a copper kingdom in Mexico's Cananea Mountains.*

cans were paid in silver pesos, and silver was worth merely half the value of gold.

In the midst of a larger labor turmoil that resulted in strikes among textile workers, miners, and other laborers across Mexico, labor agitators arrived in Cananea. They distributed a mass of literature, tried to form a union, and started a revolutionary newspaper, *The Regenerator,* which advocated "Mexico for the Mexicans" and charged that President Porfirio Diaz had sold out to the *gringos.* By noon on Friday, June 1, 1906, some 5,000 Mexicans, chanting *"¡Viva Mejico!",* had left the mines and congregated in the main plaza to demand that their wages be raised from three to five pesos per day. Waving red flags, they also insisted upon an eight-hour workday and stated that they would go on strike if their conditions were not met immediately.[6]

Within half an hour Greene addressed the angry miners from the long veranda of the Cananea Consolidated office building. Playing for time, Greene claimed that he was inclined to cooperate but that he could not grant such concessions without the consent of Sonora Governor Rafael Yzabel.

The volatile crowd did not buy it. They marched to the lumberyard to enlist the employees there, but the manager and his brother turned a high-pressure water hose on them. The strikers mobbed the two Americans, stabbing them to death with miners' candlesticks and setting the lumberyard on fire. As mine whistles began to screech the danger signal, Mexicans armed themselves with weapons seized from various stores and attacked American mine workers. William D. "Dave" Allison, a former Texas Ranger, Texas sheriff, and lieutenant of the Arizona Rangers, had been hired by Greene as a security man, and he organized the American defense, forting up at Greene's massive mansion. Men were shot on both sides, and American women and children feared that their defenders would be overwhelmed. Violent rioting raged all day, and after dark a train was loaded with Americans for a run toward Arizona.[7]

Greene frantically sought help from the outside world. Governor Yzabel and Gen. Luis Torres sped up by rail from the capital at Hermosillo. Colonel Kosterlitzky and his mounted column were fighting Yaquis far to the west, but the *Rurales* immediately commenced a forced march toward Cananea. A Friday call to Bisbee connected with Captain Rynning, in town on official business. Greene also called Douglas and El Paso for aid, while U.S. Consular Agent W. J. Galbraith, who also served as company physician in Cananea, telegraphed Secretary of State Elihu Root: "Send assistance immediately to Cananea, Sonora, Mexico. American citizens are being murdered and property dynamited. We must have help. Send answer to Naco." [8]

At 10:15 Friday night, a train from La Cananea crossed the line, rolling into the Naco station and disgorging nearly 1,000 refugees. Most of the frantic passengers were women and children, and seats had been removed so that more people could be crowded aboard. Excitedly they told their stories, reporting hundreds of casualties. These numbers later proved to be greatly exaggerated, but by now everyone was spoiling for a fight and eager to believe any lurid detail. Fifteen minutes later a two-coach train arrived with more refugees and more tales.[9]

As word spread, hundreds of men swarmed the streets of Bisbee, clamoring to invade Mexico and rescue their countrymen. That night American bordertown newspapers printed special editions to keep their agitated readers informed. Citizens wanted Rynning and his Rangers to lead a rescue expedition. Walter Dodge, manager of the Phelps-Dodge Mercantile Company, told Rynning that Billy Brophy, in charge of the company's store in Naco, would make available all of the arms and ammunition in the local warehouse.[10]

Rynning decided to act, remarking that he probably would lose his Ranger commission. He began selecting men for a relief force while a special train was put together in Bisbee. Two veteran Rangers, Arthur Hopkins and Sam Hayhurst, were on hand, and Rynning also rounded up a few ex-Rangers, as well as several men who had soldiered in Cuba or the Philippines. The train, made up of rolling stock from the El Paso Southwestern Railroad, steamed out of Bisbee at midnight. Rynning and his volunteers advanced to Naco, where they were issued arms and medical supplies. Rynning organized his men into companies under the command of experienced hands, who conducted rudimentary military drills "to get the feel of discipline into them." Armed horsemen galloped into town and were incorporated by Hopkins into the group, which finally numbered more than 250. "We had all kinds of arms," said Hayhurst, "six shooters, rifles and ammunition." [11]

All the while B. A. Packard, a banker and cattleman from Douglas who was closely associated with Greene, manned the telephone line to Cananea. Taking telegrams as well, Packard relayed news and pleas for help. One wire from Greene read: "FOR GOD'S SAKE SEND US ARMED HELP." Governor Kibbey, learning of the situation, began to telegraph

Rynning not to enter Mexico, but Packard kept these messages for delivery at a less inhibiting time.[12]

Rumors abounded that Mexicans would try to halt Rynning's relief train, and American horsemen began patrolling the line around Naco to halt any attack. Just before midnight, a thirty-man contingent tried to cross into Mexico. A skirmish broke out with Mexican riders, and two Americans were shot out of the saddle. Soon there was sporadic gunfire up and down the line, and half a dozen Mexicans were hit.[13]

Governor Yzabel and General Torres arrived in Naco on the way to Cananea early Saturday morning. Meeting these officials at the Naco line, Rynning informed the governor that his intentions were peaceful but that his men would not be halted from their mission. Rynning's tone was firm and the mood of his men menacing; Governor Yzabel realized he could not stop the Americans, and he may have felt that they could be useful in quelling the uprising in his province. Yzabel and Torres attempted to furnish a measure of legitimacy by making the Americans volunteers in the Mexican army. "I was sworn in as a colonel," reminisced Rynning.[14] Sergeant Hopkins was temporarily appointed a lieutenant colonel, and two Bisbee constables became majors.[15]

The train finally pulled out with the governor and his escort, crossed the thirty-five miles of track from Naco to Cananea, and arrived about an hour before noon. Rynning, already familiar with Cananea from previous visits, ordered a number of riflemen to deploy on a hill which commanded the area. Other "Bisbee volunteers" reinforced the Americans at Colonel Greene's mansion, where a large number of women and children had sought refuge. Exposing himself to a hostile mob, Greene again addressed his disgruntled miners, courageously standing in an open touring car in front of the Cananea Consolidated office building while many of the new arrivals backed him up, rifles at the ready. "Colonel" Rynning allowed a contingent of his men to make a quick sally to Chivatera, where the night shift had come out throwing dynamite and "the lid had blown off." [16]

The predicament calmed rapidly after the arrival of the armed force, but there still was sporadic violence. "Saw plenty of fighting," recalled Sam Hayhurst more than thirty years later. "Lots of people got killed." [17] Rynning told of hiking over to the hospital in response to a request for help from Doctor Galbraith. On the way Rynning passed a few American women sitting on the porches of their flimsy wooden houses, but when he warned them to go inside "they just laughed and said they were as safe [from bullets] outside as in." The hospital was under sniper fire, and Rynning spotted three Mexican riflemen on the bridge about 200 yards north of the building. One sniper was standing, one was kneeling, and the other was in a prone firing position. Rynning threw his rifle to his shoulder and touched off three rounds. He thought he scored a hit with each shot, and all three snipers scrambled off the bridge.[18]

At 7:00 that evening Colonel Kosterlitzky rode into Cananea at the head of a column of *Rurales*. The leaders of the insurrection fled into the

*W. C. Greene stood in an open touring car to face a hostile mob of striking miners. Behind him well-armed Americans are poised for trouble in front of the Cananea Consolidated office building.*

hills as martial law was proclaimed. Saloons were closed, *Rurales* searched everyone they encountered for guns, residents were warned to clear the streets under penalty of being shot, and several Mexicans were summarily executed during the night. Colonel Greene himself led a patrol of *Rurales* through the streets. Kosterlitzky ordered that leaders of the riot be " 'dobe-walled," or shot, and these men were laid out near a stable with their arms crossed and their *sombreros* across their faces.[19]

By Sunday morning only a few occasional gunshots suggested the turmoil of the first day of rioting. Rynning and his men were asked to leave, and at mid-morning the American volunteers escorted a number of women and children past several bodies in the street toward their train. As they pulled out of Ronquillo a few sniper bullets struck the cars, but the train continued northward. Fifteen hundred *soldados* were dispatched to Cananea from Mexico City, while four troops of U.S. Cavalry from Fort Huachuca were poised on the border. On Monday the miners prudently began returning to work.[20]

Back in Arizona, Rynning received a summons from Governor Kibbey to come to Phoenix. The captain complied, realizing that his commission was on the line. But Arizonans along the border regarded Rynning as a hero for leading the rescue operation, and the governor's anger was quickly soothed by a full explanation of the circumstances. The Rangers, after all, had gone into Mexico for reasons compelling to frontiersmen: blood called to blood. "You see," said Sam Hayhurst, "we had a lot of

*Capt. Tom Rynning (third from left) and Col. Emilio Kosterlitzky (on white horse) met at strike-torn Cananea. Rangers Sam Hayhurst (far left) and Arthur Hopkins backed up Rynning.*

friends in Cananea at the time of the strike. . . . Going down there was just a personal affair and nobody sent us." [21]

Personnel turnover during 1906 was minimal. Only nine men were enlisted, although two of them — Texan Roy Baggerly and Wayne Davis from Phoenix — resigned before the year ended. Most of the relatively few vacancies during 1906 were created by men who had joined the Rangers just the previous year: Reuben Burnett, who was discharged for drunkenness in July; Henry McPhaul; Joe McKinney; Charles McGarr; John Greenwood; and Dick Hickey and Eugene Shute, who had signed up intending to serve brief tenures. [22]

John Rhodes, a Texan who enlisted on the first day of August, was at the age of fifty-five the oldest man ever to serve as an Arizona Ranger. Rhodes was in the company for two years, and his superiors regarded him as an "excellent" Ranger. Captain Rynning was beginning to relent a bit on his requirement that all recruits be unmarried. Texans Billy Speed and Ben Olney, who enlisted on the first day of March and February, respectively, were married when they joined the Rangers.

The upper ranks of the Ranger company remained in stable hands. Tom Rynning began his fifth year as captain and Harry Wheeler, a Ranger since 1903, was still the lieutenant. But in October, Sgt. James McGee resigned, and on the last day of 1906 Sgt. Arthur Hopkins, who had presided over the headquarters desk since 1903, left the company to become undersheriff of Cochise County. [23]

Accusations of Ranger brutality continued. In August, Pvt. Charles Eperson, a Ranger since 1903, had to defend himself against a charge of cruelty while making an arrest. At Gila Bend, Eperson took into custody a young man named James Williams, who had stolen a ride on a freight train from Tucson to Los Angeles. Williams decided that he was treated with undue harshness, and he filed a complaint. He claimed that when he tried to run away, the Ranger had fired seven shots at him. His face was bruised and both eyes had been blackened by a brutal beating, and he alleged that Eperson had confined him in the Gila Bend jail without food and water for an entire day. But Lt. Harry Wheeler journeyed from Douglas to Gila Bend and conducted a local investigation which showed Williams's accusation to be groundless. When the hearing was held in Phoenix, Wheeler presented his findings and won exoneration for Eperson.[24]

Eperson's enlistment ended on September 7 and he signed up for another year. A newspaper had emphasized that he "has a good record as an officer and has succeeded in making Gila Bend a peaceful hamlet." [25] But two weeks after reenlisting, perhaps disgruntled over the charges made against him, the veteran resigned.

Lieutenant Wheeler, meantime, made his way back to Douglas with a stop at Tucson, where he was interviewed by a reporter for the *Star*. After commenting on the Eperson-Mitchell incident, the *Star* renewed speculation that Wheeler might soon succeed to the captaincy. It was thought that Tom Rynning might become sheriff of Cochise County in November, in which case Wheeler would be likely to fill the resulting vacancy of Ranger captain. But once more rumors of Rynning's resignation proved groundless, although he continued to be alert for promising opportunities.[26]

Captain Rynning was asked to inspect the San Carlos Reservation in August. Apparently, the Indian agent wanted the testimony of the prestigious Ranger leader to reinforce his pleas for a renewal of government-issued rations. The conditions on the reservation were predictably bad. "Rations are now only issued to the very old and decrepit Indians," testified Rynning to a Tucson newspaper reporter. "Those who are young and in health are expected to provide for themselves." Few jobs were available for Indian men, and the government wards thus were "forced to subsist largely on cacti fruit and acorns, which they gather all over the hills and mountains of Graham and Gila counties." [27]

Two months later Captain Rynning was on another assignment, this one crucial to the continued existence of the Rangers. The big ranchers of Arizona had been instrumental in creating the Rangers, lobbying for a band of lawmen who would concentrate on rustling as their primary target. But certain citizens continued to criticize the Rangers, vocally maintaining that the company was not worth the expense to the taxpayers. In November at least 150 ranchers gathered in Phoenix for the annual meeting of the Arizona Cattlemen's Association, held in conjunction with the Territorial Fair. The 1906 meeting was regarded as "one of the most im-

*Billy Speed and his wife in front of their Willcox home.*

portant ever held by the association," in part because Captain Rynning was scheduled to address the assemblage.

In between visits to fair events, the cattlemen convened for three days at Horner's Hall. On Thursday, November 15, the assembled ranchers heard a speech supporting government supervision of open range conditions, followed by an address on livestock sanitary conditions in Arizona. Then Tom Rynning took the platform.

The tall Ranger captain outlined the origins of the company, emphasizing that the Rangers were created primarily to rid the territory of rustlers. He described the methods used by the Rangers in the field, and pointed out that the company averaged 1,000 arrests a year, with an eighty percent conviction rate.[28] The cattlemen responded enthusiastically, adopting a resolution that the "Association fully appreciates the good work done by the Ranger force within our territory . . ., and we petition the next session of our legislature and our governor not to reduce the Ranger force at the present time." [29]

Certainly the Rangers still were needed, as evidenced by a bloody saloon holdup in the mining camp of Helvetia on June 1. Located thirty miles southeast of Tucson, Helvetia boasted 300 people, a few adobe buildings, and a cluster of tents and grass shacks. When two masked men marched into one of the saloons, a bartender slipped into a back room and returned with a Winchester, and a wild exchange of shots erupted. A Yaqui youth was fatally wounded, but the masked men shot their way outside, mounted up, and galloped out of town.

*Sgt. James McGee was elected sheriff of Pinal County after leaving the company, one of several ex-Rangers to continue in law enforcement.*

Sgt. Tip Stanford took up the trail, tracking the holdup men north along the Nogales road toward Tucson. Six miles from Tucson he lost the trail, but he rode into town anyway and enlisted the help of local officers. Descriptions of the masked men were scanty, however, and a search of Tucson proved fruitless.[30]

Willis Wood, later alleged to have been one of the holdup men, soon planned a more sophisticated robbery. Wood had several confederates, including James Alexander, a bigamist and counterfeiter, and Burt Alvord, who recently had been released from Yuma Territorial Prison. The gang apparently intended to steal a load of bullion from the King of Arizona Mine when it was brought to Mohawk for shipment, although it is possible that they intended to rob a Southern Pacific train near Mohawk. Mohawk lies fifty miles east of Yuma, and the outlaws meant to head south through the Mohawk Valley, then across the Tule Desert and into Mexico. Most of the thirty-mile ride to the border is through arid land so barren that scarcely any vegetation grows. The gang, however, went to the trouble of filling and placing twenty water barrels along their escape route. Barrels were located every five miles, and the outlaws planned to water their horses, then empty the remaining water so that any pursuers would have to proceed at a slower pace.

The plot was uncovered by Lieutenant Wheeler. Shortly after traveling to Gila Bend and Phoenix to clear up the charges against Private Eperson, Wheeler was dispatched to Yuma on official business. In Yuma he recognized and arrested James Alexander, who was wanted in Graham County for various charges. Before sending Alexander to jail at Solomonville, Wheeler extricated part of the story of the robbery scheme. He was told about the water barrels and about Wood and Alvord, although Alexander refused to divulge the identity of other gang members.

Wheeler assembled Sgt. James McGee, Lew Mickey, and Charles Eperson, then rode along the planned escape route, finding the freshly filled water barrels but no robbers. The Rangers, now reinforced by a few area officers, set a trap, but the outlaws apparently sensed danger and abandoned the scheme. After a few days the Rangers rode to the Mexican border and back, destroying the water barrels, now warped in the blazing August heat. They could find no outlaws, but the plot was successfully foiled by the Rangers.[31]

In August Governor Kibbey learned that Mexican citizens were gathering secretly in and around Douglas to organize revolutionary activities against the increasingly troubled administration of Porfirio Diaz. Kibbey ordered Captain Rynning to keep a close watch on this situation, and by September 1 a majority of the Rangers had been called to duty at headquarters in Douglas. The governor also contacted the U.S. district attorney for Arizona, who began readying deportation proceedings against any apprehended revolutionaries.[32]

At least one raid was made in Douglas against political revolutionaries. In September, Rangers and federal officers arrested twelve men and seized a large supply of explosives, flags, and printed material, apparently aborting an imminent attack planned by "La Junta" against Cananea and Naco, Sonora. Confiscated correspondence also led to the subsequent arrest of La Junta leaders in Los Angeles.[33]

Similar trouble occurred along the border a little farther to the west. Peon miners were being agitated to assault Nogales, Sonora, but this scheme was nipped in the bud by two area Rangers, Sgt. Billy Old and Pvt. John Clarke, and by Immigration Inspector J. J. Murphy. The three officers had learned of rebel recruiting efforts in the Patagonia Mountains north of Nogales. At two mining camps, Patagonia and Mowry, they arrested a trio of revolutionary ringleaders. On Sunday, September 2, the lawmen took into custody Carlos Humbert, Bruna Trevino, and Genaro Villaria; Humbert was a Frenchman, while the other two were Mexican citizens. The lawmen hustled their prisoners to the Nogales jail, creating front-page headlines in Arizona newspapers. Immediately there were threats from across the border of a raid on the jail. However, Sheriff Charles Fowler swore-in fifteen special deputies to guard the prisoners, and when the anarchists were transferred to cells in Tucson later that week the threat of trouble had dissipated.[34]

Labor Day was celebrated actively in western communities, and the program invariably included baseball or horse racing or other athletic endeavors. In Goldfield, Nevada, the 1906 Labor Day featured a prize fight between Joe Gans and Battling Nelson, which attracted enthusiastic attention all over the country. Just before the fight, a quartet of Tucson men became so exercised over the respective merits of the pugilists that a four-way brawl broke out near the depot. Sergeant Old and Private Clarke were in Tucson and saw the disturbance. They let the scrappers batter away at each other, thus expending their hostilities and making arrest easier. The four men were taken in and charged with disturbing the peace.[35]

In November, Old was at his duty station in Nogales when he learned that Francisco Zepeda, a Mexican customs officer, was suspected of murdering a girl in Tucson. Old managed to catch Zepeda across the line, then arrested him and took him to Tucson. Later in the month Lieutenant Wheeler arrested Bob Donaldson in Phoenix for running a craps game. Soon it was learned that Donaldson was wanted for assault with a deadly weapon and jailbreak in Kay County, Oklahoma, and Sheriff C. W. Worden of Kay County was notified to claim the prisoners.[36]

During 1906 the Rangers tried to deal with a recurring border problem: smuggling. In the summer of 1905 the Rangers had been alerted to halt arms sales in Arizona to Yaqui Indians from Sonora. The Yaquis long had warred against the Mexican government (it was rumored that Burt Alvord and Billy Stiles led one band of Yaqui warriors), and had found Arizona gun dealers a ready source of arms and ammunition. But by 1906 Governor Kibbey felt able to report that, due to the efforts of the Rangers and other officials along the border, the "traffic in arms . . . for the benefit of the Yaqui Indians has practically ceased." [37]

Despite Kibbey's optimism, gunrunners continued to sneak arms across the border. An officer named Fred Rankin, who rode the line out of Naco, stated that he and Jeff Kidder were receiving forty percent from the Mexican government on all the smuggled goods they found. Rankin and Kidder worked together on the border for several months, on one encounter fighting a pitched battle against a gang of smugglers. Kidder shot the arm off one of the outlaws, while Rankin managed to kill the mount of another gang member. The surviving smugglers fled, abandoning 10,000 rounds of ammunition and miscellaneous other contraband that was seized by Kidder and Rankin. The Mexican line riders and officers of Naco, stymied by the smugglers, jealously threatened the lives of Kidder and Rankin. In time they would have a chance to make good their threats.[38]

Jeff Kidder's father died in 1905, leaving Jeff a handsome inheritance. Kidder bought a new single-action Colt .45, an engraved, silver-plated weapon with pearl grips and a five-and-one-half-inch barrel. He continued his ceaseless pistol practice, firing off so many rounds that in 1907 he had to send his Colt back to the factory for replating and other repairs. Kidder came to be regarded as the quickest draw in the Ranger force. Although there were a number of crack marksmen in the company, Jeff also was considered "one of the best shots of the Rangers. At 30 steps he demonstrated repeatedly that he could hit a playing card three times

*In 1905 Jeff Kidder bought an engraved, silver-plated Colt .45 with a 5¹/₂-inch barrel. He fired so many practice rounds that in 1907 he had to send the Colt back to the factory for replating and other repairs.*

out of six on an average. He could shoot also equally well with either hand." [39]

Kidder was in Naco in 1906 when more than a dozen hobos clustered at the railroad tracks just west of town and began drinking. The disturbance was reported, and Kidder and Sgt. Bill Sparks walked out to break up the roisterers. But the tramps mobbed the lawmen and, with a large crowd of bystanders looking on, the two Rangers were pummeled to the ground.

As Kidder struggled to his feet, one of the tramps came at him from the rear, brandishing a knife. Sparks saw the threat and fired a revolver bullet which caught the assailant in the hip. At the sound of the shot the trouble ended. The miscreants were taken into custody, and the wounded knife artist was transported to the county hospital in Tombstone. [40]

On New Year's Eve, 1906, Kidder was involved in another shooting. Widespread robberies and burglaries in Douglas had stimulated the Rangers to crack down on their headquarters community. Early in December the tough Kidder was temporarily transferred from Nogales to Douglas to beef up the headquarters squad, and he regularly patrolled the border town in search of suspicious characters. [41] On the night of December 31, Kidder and a local peace officer named George Campbell were riding in the vicinity of the railroad roundhouse when they encountered a local saloonkeeper named Tom T. Woods. Woods emerged from a rear door and began to scurry through the rain across the railroad tracks.

"Hold on there," shouted Kidder, "we want to look at you."

Woods instead broke into a run, then turned and fired a pistol shot at Kidder, who was thirty or forty yards ahead of Campbell. Kidder drew his Colt and blasted out three rounds. One slug slammed into Woods's right eye, dropping him on the spot.

Kidder hurried to the side of the fallen man. Woods's gun lay three feet away, a .38 revolver on a .45 frame, and in his hip pocket Kidder found a sock filled with .38-.40 smokeless cartridges.

Woods was unconscious but still alive. Bystanders crowded to the scene, and when a physician arrived he administered restoratives to Woods. The wounded man was carried to the Calumet and Arizona Hospital, but he never regained consciousness and died that night.[42]

Kidder, by now sensitive of his reputation, requested a murder trial to obtain a full investigation in court. At his hearing on January 10, 1907, there was no testimony unfavorable to the Ranger. The prosecuting attorney moved to dismiss the defendant, and the court found that Kidder had acted "in the proper discharge of his duty." [43]

Jeff Kidder was exonerated. But he never stopped practicing with his six-gun, and he remained eager for any challenge to his person or office. In some ways Kidder was the embodiment of the Rangers: tough, quick on the trigger, ready to deal harshly with suspected lawbreakers. These qualities would be evident in the company at large in 1907.

# 1907:
# A Tighter
# Grip on
# the Force

*"Every man is a guardian of the honor and reputation, not only of himself, but the entire [Ranger] organization."*
— **Capt. Harry Wheeler**

It would prove to be a landmark year for the Arizona Rangers. During 1907 Rangers distinguished themselves in violent confrontations, displaying superlative courage and combat skills while putting their lives on the line to uphold law and order in the territory. There also was a change of command in 1907, as well as a transfer of headquarters.

The first spectacular event of 1907 occurred in February, a gunfight which became one of the classic *mano a mano* duels in all of frontier history. The gunplay occurred on Thursday, February 28, in Benson, where a lovers' triangle exploded into violence. J. A. Tracy, agent for the Helvetia Copper Company at Vail's Station, traveled thirty miles by rail east to Benson for a murderous confrontation with Mr. and "Mrs." D. W. Silverton.

Silverton, member of a prominent Kentucky family, had studied mining engineering in college and had traveled west "to see the practical side of mining." To satisfy his romantic side he was staying in Tucson with a tall, shapely brunette in her mid-twenties. The couple claimed to have been married six weeks earlier in Phoenix by an evangelist named Mc-Coma, but later no one could recall a traveling preacher by that name, and no marriage license could be located.

The woman apparently met both men in Nevada: Silverton late in

111

*There was an explosive confrontation in Benson when Harry Wheeler, Tom Rynning, and the members of a lovers' triangle converged on the town in February 1907. The depot may be seen behind railroad cars at far right.*

1905, and Tracy early the next year. Tracy evidently had been married to her (Arcus Reddoch called him "her estranged husband," and she had lived in Vail's Station with Tracy), and when he learned that she was in Tucson he paid her a visit to offer her a diamond ring. She declined, and Tracy returned to Vail's Station without comment. The next day, however, she received four threatening letters from her frustrated suitor.[1]

The couple decided to tour Douglas, Bisbee, and Cananea, and they boarded a train Wednesday afternoon. En route to Benson, where they planned to spend the night, the train made a stop at Vail's Station. Looking out of her window, "Mrs. Silverton" caught sight of Tracy standing on the depot platform. When she pointed out Tracy to Silverton, he bounded out of the car and the two men exchanged words heatedly. As the train pulled out of Vail's Station, Tracy angrily tried to catch the platform of the rear car.

An hour later the train arrived in Benson and the couple checked into Room 14 of the Virginia Hotel, across the street from the depot. Silverton engaged a porter to tell him if Tracy showed up, and early the next morning he received word that Tracy was indeed in Benson.

Tracy had come by freight train during the night, armed with a Colt .45 and muttering that he was "going to Benson to get a couple of people." When Silverton emerged onto the hotel porch on Thursday morning, he spotted Tracy standing beside the train for Bisbee. Ducking back inside, Silverton told the hotel proprietor, Eduardo Castañeda, that he needed a

*The Southern Pacific Railroad Station, site of the classic gunfight between Harry Wheeler and J. A. Tracy.*

gun for protection against Tracy. Castañeda opined that he should seek a law officer instead of procuring a firearm.

The nearest lawmen happened to be the two highest ranking officers in the Arizona Rangers, who were staying in the Virginia Hotel. Captain Rynning had recently taken a fall from a horse, badly injuring his back and hip. He had holed up in the Virginia Hotel to take massage treatments from a Benson physician and had wired Lieutenant Wheeler, who was stationed in Willcox, to take charge of the company. Wheeler arrived and checked into the hotel to receive further instructions from Rynning. Castañeda, knowing that Rynning was crippled, sent for Wheeler, who was eating breakfast.

Wheeler listened to Silverton's story and was shown a photograph of Tracy by the brunette. Wheeler searched Silverton for any concealed weapons, then headed for the depot to disarm Tracy. The man was sitting on the steps of a dining car, but as Wheeler approached, Tracy saw Silverton and his paramour come out of the hotel. Tracy jumped up cursing and pulled the revolver from his pocket.

"Hold on there," barked Wheeler. "I arrest you. Give me that gun." [2]

Relentlessly stalking his foe, Wheeler advanced, firing methodically and ordering Tracy to surrender. Tracy's third shot wounded Wheeler in the upper left thigh near the groin, but the Ranger drilled his opponent four times, tumbling him onto his back. Tracy was hit under the heart, in the neck, arm, and thigh, and he gasped, "I am all in. My gun is empty."

Wheeler dropped his Colt, having fired all five rounds, and limped forward to secure his prisoner. But Tracy had two bullets left, and he treacherously opened fire again, striking Wheeler in the left heel. Gamely, Wheeler began hurling rocks at Tracy, whose gun finally clicked on an empty cylinder.

"I am all in," he repeated. "My gun is empty." [3]

Wheeler limped to him, but Tracy refused to give up his gun (later several cartridges were found in his pocket). At that point the porter ran up flourishing a small pistol, which Silverton tried to seize from him — both men intended to pump more lead into Tracy.

The bleeding Wheeler managed to calm them as other onlookers crowded around. Wheeler finally disarmed Tracy and someone brought a chair for the wounded officer.

"Give it to him," said Wheeler, "he needs it more than I do."

Wheeler relinquished responsibility for Tracy to a Benson peace officer, then extended his right hand. "Well," said Wheeler, "it was a great fight while it lasted, wasn't it, old man?"

"I'll get you yet," muttered Tracy with a hint of a smile.

The two men shook hands, then Wheeler retrieved his own six-gun and, with assistance, headed for the hotel. Captain Rynning hobbled outside and took a statement from Tracy. The scorned suitor asked to be sent to a Tucson hospital, and he was placed on a cot in the baggage car. But by the time the train reached Mescal Station, just ten miles west of Benson, the thirty-eight-year-old Tracy had breathed his last. His final words were: "There is a woman in the case." His remains were shipped to Chicago.[4]

A reporter from the Bisbee *Review* asked for a statement from Wheeler, who replied that he was sorry Tracy had died. Rynning's doctor tended Wheeler's wounds and suggested he seek treatment at the hospital in Tombstone. Rynning accompanied his lieutenant by rail thirty-five miles to Tombstone, where Wheeler was hospitalized for two weeks. But the wound near his groin never ceased to trouble him, especially on long horseback rides.

In June it was learned that J. A. Tracy had been wanted for two separate murders in Nevada, with a $500 reward on his head. One of his victims was the brother of former Ranger Dick Hickey, now a justice of the peace in Pinal County. Nevada officials offered Wheeler the reward, but he immediately turned it down. Although he was "a poor man," Wheeler would have no part of blood money, instead urging that the $500 be given to the widowed Mrs. Hickey.[5]

During the early months of 1907, a gang of horse thieves was working the border, rustling stock and driving the animals in and out of Mexico. One of the rustlers, a young Mexican, was pursued by Sonoran officials into Cochise County near Douglas. The Mexican officers contacted Arizona authorities, and soon Rangers Sam Hayhurst and Jess Rollins were on the trail.

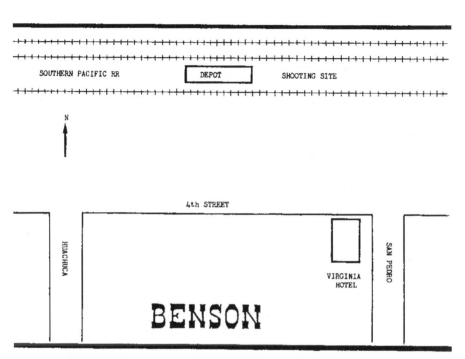

SOUTHERN PACIFIC RR       DEPOT       SHOOTING SITE

N

4th STREET

HUACHUCA

VIRGINIA
HOTEL

SAN PEDRO

BENSON

A tip sent them to a stone quarry east of Douglas, where numerous hovels were located. Inquiring at a tent, Hayhurst and Rollins encountered an attractive young woman who proved to be the wife of their prey. She told the Rangers that the man they sought was not there, but Hayhurst and Rollins skeptically searched the area. After a few minutes they found their man hidden by several cots and a pile of quilts. The Rangers dragged the rustler from his hiding place as his wife pleaded for his release.

Jailed in Douglas until extradition procedures could be performed, the fugitive expressed "deep regret at being in trouble" — as well he might, since horse thieves usually were shot in Sonora. This was one of 614 arrests made by the Rangers during the year, a routine apprehension that was typical of the activities which were noticeably taming Arizona Territory.[6]

In March 1907 Tom Rynning finally received the promotion he long had sought. Jerry Millay, superintendent of the Yuma Territorial Prison, resigned because of ill health, and the position was offered to Rynning. "Of course it was some wrench to leave the Rangers," reminisced Rynning, who had captained the company for all but the first year of its five and one-half years of existence. But Rynning considered the hardest work of the Rangers to have been accomplished, and he was a politically oriented man with administrative gifts who was eager for a fresh challenge. A new prison was being established at Florence, and Rynning, who had been a contractor prior to his Ranger service, eventually was placed in charge of construction at Florence, using convict labor to build the complex. Rynning would conduct the move from the adobe pile along the Colorado to the new facility, where he intended "to make decent citizens" out of many of the same lawbreakers his Rangers had apprehended. Rynning resigned his commission on March 20, 1907, and ten days later he arrived in Yuma.[7]

Each time Rynning's resignation had been rumored, it had been speculated that Harry Wheeler would be promoted to captain. There was considerable public interest about who would become the new captain of the Rangers, but any suspense over the matter would be of brief duration. Harry Wheeler had impressed everyone with his bravery and skills as an officer and with his intense devotion to duty. One supporter, Sheriff John White of Cochise County, wrote Governor Kibbey praising Wheeler in ways typical of those who knew the man: "He is one of the finest, and most honorable gentlemen that it has ever been my good fortune to know, his character is without blemish. . . . As a peace officer Mr. Wheeler is most competent, . . . highly intelligent, a man of splendid judgement, cool — skillful, daring, and the right man in the right place at all times." [8]

Governor Kibbey promptly contacted Wheeler about the position, and Harry just as promptly accepted. Indeed, Arizona newspapers had predicted Wheeler's promotion even as his gunfight with J. A. Tracy made front-page headlines. The legislature confirmed his appointment on March 22, and on Saturday, March 25, Harry Wheeler took the oath as third captain of the Arizona Rangers.[9]

Two days later Sgt. Billy Old was promoted to lieutenant. His close friend, Sgt. Jeff Kidder, had been mentioned for the lieutenancy. Both men were highly capable officers and crack shots, but Kidder, despite a year's seniority over Old as a Ranger, used the authority of his badge with a heavy hand and seemed to attract trouble. Billy Old proved to be a diplomatic and capable choice as the Ranger second-in-command. Old had been stationed for years in Nogales, but Wheeler sent him to Prescott so that he could direct Ranger activities in the northern part of the territory.

One of Wheeler's first moves was to transfer headquarters from Douglas to Naco. Douglas had built up quickly, and the Ranger presence there had exerted a taming influence. The Phelps-Dodge Company had made a determined effort to improve the moral climate of the town; for example, so much money had been spent — much of it by the company — on religious edifices that Douglas now was called "the City of Churches." [10] But twenty-five miles to the west, just across the border from Naco, Sonora, the raw American community of Naco, Arizona, had become a pesthole.

Just after the turn of the century, the Phelps-Dodge Company considered building a smelter at Naco. When the owners of the townsite refused to donate land, the smelter was erected at Douglas, and Naco was con-

*In March 1907 Capt. Tom Rynning was appointed to the superintendency of Yuma Territorial Prison.*

demned to a secondary existence. But W. C. Greene owned a residence at Naco, and the freight activity stimulated by his massive operation at Cananea caused a certain amount of economic progress. Freight cars loaded with mining equipment and supplies rumbled south through the big opening in the border fence, and great slabs of copper were shipped north through Naco. A brick bank was constructed on the 100-foot-wide main thoroughfare, "D Street," which runs north and south. The Copper Queen mercantile building also was brick, and the Hotel Naco was a two-story adobe which still looms on the east side of the street. Greene kept a dozen employees busy in an office building, several warehouses lined the railroad tracks, and a $30,000 customshouse was erected. John Baker moved his Owl Restaurant from Benson to Naco on a flat car. The Naco *Budget* offered a daily diet of boosterism to newspaper readers. For years trained hounds had coursed at the Naco Amusement Park. Naco even had a baseball team; so many nationalities were represented on the roster that one Arizona newspaper termed them "a Congress of Nations." On a trip to Tucson, Wheeler, aware that Tuscon was a baseball hotbed, bragged on the Naco club. He tauntingly claimed that the Naco manager would not allow his team to play in Tucson, since such a match "would only be a practice game for the Naco crowd." [11]

The population of Naco, Arizona, was only 500 or so when Harry

*The restaurant at the Hotel Naco.*
(Courtesy of John Payne, Bisbee, Arizona)

Wheeler brought Ranger headquarters to town. Naco, Sonora, had more people, and the nearby presence of Mexican dives added to the potential for outlawry on the Arizona side of the border. Located on the major route from Cananea to Bisbee, the Arizona border town became the scene of massive smuggling activity and a general congregating point for men of criminal inclinations. Local officials were in league with outlaws, and acts of violence were commonplace. By 1907 Naco was a corrupt community crying for the permanent presence of the Arizona Rangers.

Captain Wheeler imprinted a forceful stamp on the men of the Arizona Rangers. Wheeler had been in the company for nearly four years, and he was the only Ranger who served at every rank — private, sergeant, lieutenant, and captain. Product of a nineteenth-century military upbringing, he brought patriotic fervor as well as up-from-the-ranks experience to the command post of the Rangers. He came to the captaincy with strong ideas about the mission and methods of the company, and with an iron will he endeavored to compel each Ranger to meet the highest standards. On June 1 Captain Wheeler issued a set of General Orders which revealed to the men under his command his exacting expectations:

GENERAL ORDER #1.

Rangers will not congregate in saloons; nor in any bawdy house, for the purpose of amusement, or out of idleness.

This is not intended to interfere with any Ranger acting in his offi-

*The United States customshouse at Naco.*
(Courtesy of John Payne, Bisbee, Arizona)

cial capacity; nor to prevent his taking a drink if so inclined; nor the exercise of any right or privilege in which any gentleman would feel safe.

### GENERAL ORDER #2.

It being a well established fact that men of honor, will meet their just, financial obligations on time and according to promise, the attention of all Rangers is called to the evil of going beyond their means, and for any reason borrowing or contracting loans they will be unable to meet.

The reputation of this company has been *very* high, but of late one or two incidents, which have come to my notice, have caused me much embarrassment and humiliation. All Rangers must meet their obligations.

Every man is a guardian of the honor and reputation, not only of himself, but the entire organization. It is every man's duty to aid in perpetuating the integrity of the service, by unmasking any undesirable member, who by accident may be among us.

### GENERAL ORDER #3.

Every man is hereby prohibited from entering Mexico in any *official capacity* whatsoever, armed or unarmed. Unless by special request or permission of the Mexican Authorities, and being accompanied by some Mexican Official. Even under such conditions, Rangers will remember they go *unofficially*, and simply as any other private citizen.

### GENERAL ORDER #4.

All Rangers will enforce strictly and impartially, the laws which prevent gambling, women and minors in saloons, etc. All laws will be enforced, but I desire extreme vigilance and precaution upon the part of Rangers, in preventing the violation of the laws above mentioned.

We all know that Gambling has caused more suffering and crime, serious crime, than any other dozen causes combined. In my opinion, the complete suppression of this evil, will result in a falling off, of a least 50% of the arrests for Felonies. Consequently lessening the needs of hundreds of women and children, who have known want and privation heretofore.

I consider it an honor to any man, or organization, who aids in suppressing this evil.

GENERAL ORDER #5.

In all cases where prisoners are taken, the greatest humanity will be shown, and each Ranger is instructed to mentally place himself in the others place and act, legally, accordingly.

I want no man to needlessly endanger his life by taking foolish chances with desperate criminals, and if any one must be hurt, I do not want it to be the Ranger, at the same time I want every precaution taken to insure a peaceful arrest, and in all cases, when an arrest has been consumated [sic] the prisoner must be shown the courtesy due an unfortunate, and the kindness a helpless man deserves and gets from a brave officer.

GENERAL ORDER #6.

The new monthly reports will be made out the first day of each new month and sent in to Headquarters. Every item must be filled out and the report certified to on the back.

In addition that I may keep in constant touch with all members, a postal card will be sent in to Headquarters, each Saturday, telling where you are and where you may be reached for the next few days, as nearly as possible.[12]

The word "honor" recurs throughout Wheeler's directives, along with other terms meaningful to the man: "duty," "integrity," "gentleman." The company would war against gambling, shun saloons, and avoid personal indebtedness. Scrupulously, Wheeler insisted that his men avoid the casual entries into Mexico that had been common throughout the Rangers' existence. Within a few days of these general orders, Wheeler and some of his men chased four fugitives across the border, but the captain would not allow pursuit south of the line.[13]

The Ranger company was composed of men whose characters contained strong doses of frontier individualism, and many of them resented new regulations which smacked of puritanism and bureaucratic red tape. There was to be a new precision to arrest reports, and a *weekly* postcard report to headquarters. "Of course, some men are high spirited," commented Wheeler, "to the extent that any restriction whatever, placed upon them causes them to feel injured."[14] Many of the tough, pragmatic adventurers who made up the company felt injured, and a clash with their idealistic, strong-willed new captain was inevitable.

Wheeler explained to Governor Kibbey: "I do not wish to trample upon the liberties of the men under me, nor take advantage of my position in any way," but he regarded these orders as necessary "for the good of the service in general." It should not have surprised Wheeler that the new mandates "have met cordial reception by some of the men, but I intend they must be obeyed, even to the extent if necessary, of letting those go, who feel they are unable to abide by them." [15]

True to his word, Wheeler began "letting those go" who would not comply with his Ranger ideal, and during his tenure as captain he discharged numerous men for one transgression or another. Throughout the history of the force there were enlistees who served only briefly before deciding they did not like the service, and each captain was forced to deal with recruits who demonstrated undesirable qualities. While the turnover rate under Wheeler was only slightly higher than the rate under Captains Mossman and Rynning, there can be no doubt that Wheeler exerted a stern pressure upon his men. During Wheeler's less than two years as captain, twenty-six men left the force: many resigned for undetermined reasons ("I wish a few more would quit," remarked Wheeler after weeding out one or two undesirables),[16] but several were forced out because of drinking problems or insubordination.

In 1907 the Rangers enlisted fourteen men, including nine who hailed from Texas. Four had previously served as peace officers, while seven listed "stockman" as their occupation. Frank A. Ford, at twenty-two the youngest recruit of the year, listed his occupation as stenographer; he served as headquarters clerk for less than four months before he resigned on the first day of August. W. F. Bates, a July recruit, also lasted not quite four months before Wheeler decided that he was unsatisfactory as a Ranger. George L. Mayer enlisted in August and resigned three months later, and in November Travis Poole, another August recruit, was discharged for "leaving Territory without permission." James Smith signed up in June, but despite an "Excellent" rating by Wheeler he chose to leave the company in October.[17]

Late in 1907 James Emett, a rancher from Lees Ferry near the Utah border, applied for a Ranger position, using his personal influence with Territorial Attorney General E. S. Clark and circulating a petition to assure his appointment. But information soon came to the governor's office that Emett wanted a Ranger badge solely as a license to pursue a feud with the cowboys on a neighboring ranch owned by B. F. Saunders. It was discovered that Emett had a shady reputation; he had been defended in Utah by Clark in a case involving embezzlement and suspected homicide, and it was expected that if Emett became a Ranger he would try to kill Saunders's foreman. The controversy over Emett doomed his chances, and Wheeler never accepted him into the company.[18]

Wheeler did not continue Rynning's practice of enlisting former Rough Riders as Rangers. He found "Teddy's fighters" inadequate for his purposes "because not one soldier in 100 can either rope or read a brand, and Rangers must have these accomplishments . . . as their duties are most frequently along the lines of the livestock business." [19]

Wheeler's most interesting recruit of 1907 was William A. Larn, a thirty-three-year-old cattleman originally from Fort Griffin, Texas. Larn signed his enlistment papers on October 1, 1907, in Williams, but his enrollment as a Ranger was not made public. Larn was to work undercover to break up a band of cattle rustlers in northern Arizona. The gang had "successfully defied all efforts . . . to capture them for years," but Lieutenant Old expected to land the thieves with Larn's help. "We expect great things from this man Larn," exulted Wheeler to the governor's secretary, adding, "no one at all knows that he is a Ranger." [20]

Great things from Larn did not materialize, however. Ranger arrest records do not indicate the incarceration of any rustlers in northern Arizona during this period, and on the last day of the year Larn resigned. The reasons were vague, but on his personnel file Wheeler wrote, "Service unsatisfactory." [21]

A few Rangers had been designated "Inspectors." This was never a rank established officially by the legislature, but rather an informal designation of Rangers who worked regularly as livestock inspectors. The role of inspector had evolved under Rynning, but Captain Wheeler decided that inspectors should perform all functions of a Ranger. Soon he found it necessary to write a critical letter to Pvt. Ben Olney, a thirty-five-year-old Texan who lived near Safford and who had enlisted in February 1906.

Olney responded with a mixture of resentment and repentance: "I will say that I was very angry when I read [your letter] and I felt pretty sore for I never knew I was to go every time . . . a Mexican killed another in Clifton or Morenci or Globe. . . . My understanding when I first enlisted under Mr. Rynning was that I did not have much to do being an Inspector, but . . . if I have not done enough to suit you I am sorry and I am willing to do better." Olney was married and emphasized to Wheeler: "I would like to hold the job if I can because I need the money. . . . I have worked for wages nearly all my life and I never got discharged yet. I have always been a good friend to you and all the Boys and I don't know of any ranger in the company now that I don't like."

Olney, who lived more than three miles outside town, told Wheeler that he would have a telephone placed in his home so that he would be more accessible, and he outlined a program of law enforcement activities he hoped to pursue if permitted. Wheeler decided to give Olney, whom he regarded as "very honest," another chance, and Ben's work proved satisfactory. Although he never saw any action as a Ranger, he served until the company was disbanded.[22]

Another Texan, Porter McDonald, had enlisted in June 1905 and was designated an inspector by Rynning. McDonald worked out of Tombstone, doubling as a Cochise County deputy sheriff under Sheriff John White. Although Rangers were prohibited from serving in other law enforcement capacities, McDonald accepted fees from Cochise County. He also used a railroad mileage book given him by Sheriff White, which permitted the holder to free passage. Governor Kibbey had directed Wheeler to take up any such books held by Rangers. Everyone complied except McDonald, who wrote Wheeler protesting that White had given him the mileage book: "I did not get it from the assistance of the Ranger Force or by request of the Gov. nor by a request of my own." Threatening to send Wheeler *"the star"* upon request, McDonald defiantly stated: "I absolutely refuse to turn this mileage book over to any one except those that gave it to me." [23]

Wheeler, of course, was furious when he read McDonald's letter. The captain immediately wired McDonald to accept his resignation, which was dated November 26, and he forwarded the offensive letter from "ex-Ranger" McDonald to the governor. Wheeler explained the situation involving Ben Olney as well as McDonald, pointing out that "they have the idea that when they are Inspectors, that they are removed from all jurisdiction of the Ranger Company." Wheeler commented that McDonald "has peculiar ideas and is insubordinate and the only way to get along with him is to let him do as he pleases; I have a couple more in the Company who are troublesome at times, but I am going to be the Captain. . . ." Wheeler asked Sgt. Lew Mickey to assume McDonald's livestock inspecting activities, in addition to his other duties. There would be no more "special" Rangers. [24]

Not only was Harry Wheeler a firm leader of men and an expert field officer, he rapidly proved to be a meticulous master of office detail, corresponding voluminously with his men and the governor's office. Indeed, Governor Kibbey specifically praised his administrative qualities, concluding: "Rarely has an officer been so efficient in both field and office work." [25]

After Wheeler submitted his first annual report, to his obvious chagrin he learned that at least 100 arrests had not been recorded on paper. Inquiry among his men revealed that many reports had been lost, and Wheeler instructed his lieutenants and sergeants to try to compile a list of unrecorded arrests his men could recall. He also took steps to insure that all future arrests would be precisely documented. [26]

Wheeler himself handled an unusual 1907 arrest which gave the stern Ranger leader a rare moment of amusement. On Saturday, May 18, Wheeler was in the countryside near Naco when he was approached by two deserters from Fort Huachuca. "We have avoided Naco," they confided "because of the d — d rangers." They asked the exact location of the Mexican border, which Wheeler pointed out while they continued to curse the Rangers. Then Wheeler, enjoying himself hugely, politely mentioned that they could not proceed into Mexico. When they asked why, "I told them that they had avoided the rangers alright but had come to the rangers Captain for information." Wheeler, of course, took the pair into custody and held them for the military authorities. [27]

Sgt. Jeff Kidder, as usual, remained active in seizing wanted men. On Friday night, November 22, Kidder overtook four thieves at Benson. Loaded with watches, jewelry, guns, and other plunder, the outlaws were surreptitiously boarding a freight train at the Benson depot. Although accompanied only by Constable Page, the Ranger sergeant charged the badmen and a free-for-all broke out. Kidder refrained from using his six-gun even when one of the thieves jammed a revolver into Page's stomach. Kidder grappled with the gunman, who fought furiously against the officers. During the scuffle two of the thieves managed to slip away into the darkness, but Kidder and Page subdued the would-be gunman and one of his confederates. A local policeman named Banta witnessed the end of the struggle, commenting that "Ranger Kidder showed the coolest judgment and best nerve combined of any officer he ever saw." [28]

The next month, at Huachuca Station, Kidder nabbed horse thief Thomas Larrieau. Wheeler wrote his congratulations, although Kidder's "valuable captive" regained his freedom when the grand jury dismissed the case.[29] Overall, however, the Rangers did not too frequently suffer the lawman's classic exasperation of making an arrest only to see the malefactor released. In December, for instance, the Rangers recorded thirty-eight misdemeanor arrests: one case was dismissed by the plaintiff, one was acquitted, and the other thirty-six offenders were convicted.

In August, Captain Wheeler reported a total of just twenty-seven arrests — felonies and misdemeanors combined — by the Rangers, but travel reached 9,000 miles. The Rangers rode 7,500 miles on horseback, more than eleven miles per man per day, as well as 1,500 miles by rail. In November the Rangers rode 8,493 miles, more than twelve miles per man daily. Wheeler had directed his men to change emphasis and disregard arrest totals: "the men have been instructed to do more riding . . ., and pay more attention to the serious crimes and less to the small trivial things easily taken care of by the local officers. This policy is sure to improve the conditions of any section where Rangers are stationed, constantly riding the country in sections not usually travelled by other officers." [30]

Wheeler preferred, wherever possible, to station his men in pairs. Chapo Beaty recalled that "Rangers always traveled by two's," a simple precaution followed by modern policemen who cruise in pairs in patrol cars. Aside from the obvious safety factor, Wheeler knew that two Rangers could corroborate each other's testimony on a witness stand.[31]

Captain Wheeler felt that a machine gun would prove useful during strikes or other situations which promised mob action. He requested a .30-caliber Colt machine gun, but apparently the governor, sensitive to complaints about the expense of the company, never authorized purchase of such a weapon for the Rangers.[32]

In a direct move to keep expenses down, the governor decided to maintain the roster below authorized strength. Earlier in the year there had been a strong movement in the legislature to reduce the company by more than half. In February a bill was defeated to abolish the company, but Ranger opponents countered by pushing through committee an amendment to the existing Ranger bill that would limit the force to a captain, a lieutenant, and ten privates. Although this resolution was tabled by

*Sergeants Jeff Kidder (right) and Rye Miles, bound for field duty with mounts and packhorses.*

a convincing vote a week later, Governor Kibbey seemed to feel that Ranger opponents might be appeased if the roster and payroll were reduced. As vacancies occurred during the year the departing men were not replaced, so that by June there were just twenty men in the company, six fewer than maximum strength. Numerous requests for Ranger assistance from across the territory caused the addition of a couple of men, but by the end of the year the roster totaled only twenty-two Rangers. "They work in the utmost harmony with the sheriffs and other peace officers of the several counties," optimistically stated the governor at the close of his report to the secretary of the interior.[33]

Captain Wheeler did not flinch from enforcing unpopular laws, which found passage during this period as Arizona's response to the Progressive reform movement. One of those laws involved rodeos.

Arizonans were enthusiastic rodeo fans: Prescott claims to have the oldest *continuous* annual rodeo (commencing in 1888), and Payson claims the oldest *consecutive* annual rodeo (early 1880s) — despite rival claims from such western communities as Pecos, Texas (1884) and Caldwell, Kansas (1885). By the early twentieth century, rodeos and wild west shows were popular throughout Arizona, but in 1907 the Twenty-fourth Legislature — "in response to a great wave of reform that swept before it licensed gambling and kindred immoral practices" — passed an act to prevent "Steer Tying Contests." [34]

Clay McGonagill, O. C. Nations, and Bill Pickett, famous early-day rodeo performers, had been touring the territory, astounding large crowds with feats of bronc busting, roping, and Pickett's specialty, "steer throwing." Pickett, a black cowboy from Texas known to rodeo fans as the "Dusky Demon," would gallop alongside a running steer, jump onto its back, and wrestle its head upward by grasping the horns. Pickett then would sink his teeth into the steer's upper lip and throw the animal easily. Bulldogs, when working cattle, controlled the beasts by biting their lips, and Pickett's technique became known among the rodeo crowd as "bulldogging" — the only event in modern rodeoing that can be traced back to a specific individual. An Arizona newspaper commented huffily: "To the morbid this has proven a most interesting feat and crowds have gathered expressly to see this part of the performance." [35]

The spoilsport law against steer tying contests went into effect on April 1, 1907, but for a couple of weeks Pickett and his fellow performers continued to tour Arizona with no interference. On Saturday, April 12, there was a performance at Don Luis, just south of Bisbee. A large number of "the morbid" gathered, but someone complained. Another performance was scheduled for Sunday at Don Luis, but Capt. Harry Wheeler came up from Naco and, assisted by several other officers, halted the fun by threatening to arrest anyone who threw or roped any steers. The management had the foresight to gather "a large number of unbroken bronchos," and the crowd had to be satisfied with an exhibition of broncho riding. The public had been protected from "cruelty to animals in the guise of feats of skill," and the Bisbee *Review* smugly concluded: "There will be no more of it wherever there is an Arizona Ranger." [36]

In 1907 the Rangers also were forced to clamp down on gambling, certainly a popular pastime for many Arizonans. For years there had been a movement by "the better class of citizens" in the territory to do away with professional gamblers "and women who, while not always directly associated with the gamblers, are by that far reaching term 'sports,' indissolubly linked with the men of the green table." [37]

The legislature finally decided to curb gambling activities by prohibiting all of the traditional games of chance, except poker, in public places, and saloons were to be cleaned up by forbidding the presence of women and minors in barrooms. The new laws immediately caused a rash of enforcement problems for Rangers throughout Arizona. In Nogales, for example, Jeff Kidder reported that the "saloon men here seem to think the women and minors in saloons Law is all a joke." He made one arrest in a saloon where a thirteen-year-old boy worked, and while talking with a local justice of the peace in another saloon the judge himself called a twelve-year-old bootblack inside to shine his shoes. "It has always been a custom for minors to hang around in saloons in this town," complained the exasperated Kidder.

One Nogales "resort" was operated by three women who served liquor in one room. It was customary in Nogales for women to eat at saloon lunch counters, and the prosecuting attorney feared that he would have to prosecute the wives of his friends. Saloonkeepers in Douglas, angry that they could not even employ female singers to attract customers, began collecting funds to finance a legal challenge.[38]

By August, Wheeler was able to report to the governor that public gambling "is not indulged in anywhere," except for a few underground faro and roulette games that the Rangers would have "one chance in ten thousand" of finding. In Douglas he could not even locate a craps game. But Wheeler pointed out that there was widespread abuse of the poker exemption; since poker could be played with no obvious percentage going to a house gambler, men congregated in the most popular saloons and legally played poker among themselves. "There is no doubt the law is abused in this respect," grumbled Wheeler, who was certain that a percentage rakeoff still was practiced, "but in such manner that no conviction could be obtained." He pointed out that unless the "Legislature passes an act forbidding *all Gambling* as in Texas, these lesser abuses will continue."[39] The recent Texas law had caused an exodus of professional gamblers from the Lone Star State; many El Paso high rollers had migrated to southeastern Arizona.

Frank Wheeler complained to his captain in November that in Yuma "the Women are increasing in number and the Poker Games are getting bolder and Gaining in Number." The district attorney of Yuma County, Peter Robertson, had requested special help from a Ranger in gambling matters, but Wheeler felt "he wants me to make a fool of myself." Also in November, Privates Cy Byrne and Owen Wilson arrested a woman for drinking in a Winslow saloon, but they lost this and two similar cases due to efforts of the mayor, constable, and justice of the peace. Captain Wheeler was appalled that the justice who handled the case said "he had a mind to arrest our men for carrying a gun — Said Rangers had no right to be in town & make arrests nor even carry a gun." In his December report Wheeler pointed out to the governor "that in every case of an arrest for 'Women' being in Sallons [*sic*], the Defendant comes clear." Despite varying degrees of noncompliance, Captain Wheeler still was satisfied that the curtailment of gambling activities "has had a great tendency to lessen crime of all degrees, serious or otherwise, I should say, at least 40% if not more."[40]

There was as much rough masculine camaraderie among the Arizona Rangers as among any group of males who shared adventure and danger. When Ranger headquarters still was located in Douglas, a big circus came to town. Joe Pearce and four companions "were feeling hilarious" — partially because they had imbibed "a little rat poison" — and they decided to join the customary parade through town. They invested fifteen cents apiece in costume masks, rubbed flour on their hands, and mounted up to catch the parade as it progressed through a waiting crowd on Main Street.

The parade was led by a big wagon with the band and clowns perched on top. The fun-seeking Rangers galloped their horses past the circus wagons and teams until they were in front of the big bandwagon. Since the circus was hurrying to reach the expectant crowd, the Rangers perversely reined in their mounts to a leisurely walk.

As the parade slowed to a crawl, the clowns and musicians began shouting curses at the Rangers, who shouted back and laughed heartily. Two clowns jumped off the wagon and ducked into a grocery store, but the Rangers thought they were after something to eat. By the time the parade turned into Main Street the clowns had climbed back onto their wagon, bulging paper sacks clutched in their hands.

The Rangers merrily led the circus through the crowd, but after half a block the clowns passed their bags around. Suddenly, a barrage of eggs came from the lead wagon toward the Rangers — "and they could throw about as straight as we could shoot." One egg after another splattered the Rangers' hats, clothes, saddle gear, horses' backs and, of course, dripped down their necks. The onlookers, thinking it was all part of the show, roared their enthusiasm. But the Rangers had their fill of show business. They applied quirts and spurs and raced away, followed by the laughter of the crowd and circus members.[41]

A more serious Ranger incident involved a noted rabblerouser of the day. Mother Jones was a chubby, white-haired older woman with spectacles and chipmunk cheeks — and a crafty, determined glint in her eyes that suggested why she was called "the most dangerous woman in America." Since the 1870s Mary Harris Jones had ventured around the United States making socialist harangues at coal mines, train yards, factories, and logging camps on behalf of the growing labor movement. In 1907 strikes and anti-union repression brought the controversial seventy-five-year-old agitator to Arizona Territory.[42]

Arriving in Arizona early in the year, she spent time with laborers in Globe, Bisbee, and other mining camps. Late in June "The Miners' Angel" established headquarters at a hotel in Douglas and lined up a series of meetings.[43] On Sunday, June 30, she was introduced to Manuel Sarabia, outspoken opponent of President Porfirio Diaz and a leader of the revolutionary *Junta Liberal Mejicana*. Sarabia, a small man (five-foot-three, 130 pounds) in his mid-twenties, sported a black mustache and was a natty dresser. Charged in Mexico with inciting revolution, he fled to Texas in 1904, eventually making his way to Chicago, St. Louis, and other points. About June 1, 1907, he arrived in Douglas, hiring on at the *International-American* under the alias Sam Moret. Throughout his travels Sarabia had continued long-range insurrectionary activities against the heavy-handed *presidente,* and when he met Mother Jones he boasted that he had fought Porfirio Diaz and was in the United States seeking refuge.

That night, while Mother Jones was addressing a crowd of smelter workers gathered in the street, Sarabia, who for some reason did not attend the meeting, walked to the depot on the west side of town intending to mail a letter. But on the way, Ranger Sam Hayhurst leveled a gun at Sarabia and informed him that he was under arrest.

The previous day, while traveling by train from Bisbee to Douglas, Hayhurst and Captain Wheeler encountered Ramos Bareras, captain of a newly organized Mexican border force. Bareras told the Rangers about Sarabia, claiming that he was a murderer as well as a revolutionary, and giving Wheeler Sarabia's description and Douglas address. In the customary spirit of cooperation between border officers, Wheeler agreed to arrest Sarabia and hold him until extradition papers were received.

After apprehending Sarabia, Hayhurst took his prisoner to city hall. Local officials held Sarabia incommunicado through the next day, then spirited him away into Mexico in an automobile on the first night of July. By July 4 he was in Hermosillo, where he was jailed for several days.[44]

When Mother Jones learned that Sarabia had been arrested and secretly returned to Mexico, she immediately responded to the plight of her fellow agitator. A mass meeting was held in Douglas, the resignation of the Mexican consul at Douglas (who had aided in abducting Sarabia) was called for, and complaints were filed against Hayhurst and four other officers. More than a hundred telegrams were fired off to Washington, and Arizona officials also were bombarded with protests and demands for intervention. Governor Kibbey notified Harry Wheeler to do what he could to resolve the Sarabia "kidnaping," which was attracting a great deal of adverse criticism. Mother Jones was impressed when she met Wheeler; she regarded him as "a pretty fine fellow to be captain." [45]

Wheeler contacted Gen. Rafael Torres, and both men journeyed to Hermosillo. On July 11 Torres secured the release of Sarabia, who traveled by rail back to Nogales, accompanied by Wheeler. Sarabia soon found himself back in a cell, however, incarcerated in Tucson's Pima County jail alongside three compatriots. All four men were charged with violation of United States neutrality laws. They were convicted in May 1909 and sentenced to eighteen months in the territorial prison, but by then Sarabia had jumped bail and had disappeared, rumored to have fled to Canada.[46]

Billy Old's principal assignment during 1907 was to clean out northern Arizona above the Colorado River, the wildest part of the territory. Cattle and horse thieves had run rampant south of the Utah border, primarily because the Colorado River and its Grand Canyon provided a protective barrier from lawmen. "Owing to the difficulty in getting across the canyon of the Colorado," commented a Tucson newspaper, "the outlaws have practically run things as they pleased and might have been practically the only law in that section." [47]

When Rangers had been sent to this trouble spot in the past, they had been forced to ride several hundred miles out of the way, passing through Nevada and Utah before they could reach the area. Now Lieutenant Old was stationed in Prescott to organize the "Northern Detachment of Arizona Rangers." Old had only five men in his detachment, although Jeff

*Cy Byrne, shown here long after his Ranger days, was a member of Lieutenant Old's Northern Detachment.*

Kidder joined them once to offer his formidable services. Old soon transferred detachment headquarters to Flagstaff, which was still on the wrong side of the canyon. The members of his detachment were scattered at stations across northern Arizona, but he kept at least one man above the Colorado: Pvt. W. N. Wilson, for example, was stationed for a long time at Fredonia, just below the Utah line.[48]

Most of the Rangers were desert rats who found it difficult adjusting to the wintry climate of northern Arizona. Snow can be seen at any time of the year in the San Francisco Mountains north of Flagstaff, and sleet and snowstorms begin early in the fall. The Rangers from southern Arizona quickly announced that they did not want to stay in the north, and when Wheeler told them they could remain where they were assigned or resign, four men quit the force. Old returned south in October to testify before a grand jury in Tombstone concerning arrests he had made prior to his transfer. "It is certainly mighty fine to be back in the lowlands," he cheerfully admitted, "where it is not necessary to be bundled up in heavy woolens and in overcoats."[49]

During August, Lieutenant Old organized a thorough sweep through upper Arizona. Old and E. S. McGee would scout Mohave County all the way to the Utah line, while Cy Byrne, recruit George Mayer, and W. N. Wilson would probe the isolated reaches of vast Cococino County. The Rangers intended to stay in the field four to six weeks, familiarizing themselves with the wild country and rounding up any stray badmen they could find. It was the beginning of a systematic campaign to clear the northern ranges of rustling.

The southwest corner of Arizona was capably policed for the Rangers by Frank Wheeler (no relation to Harry Wheeler). Wheeler had been a Ranger since 1902. The Mississippi native had ventured west to be a cowboy, but at the age of thirty-one he was attracted to the Rangers. A year after enlisting he was promoted to sergeant, but soon he reverted to the rank of private. He may have requested the reduction in grade: some men are uncomfortable with authority, and Wheeler, always respected as a Ranger, reenlisted year after year with no sign of resentment over his reduction in rank.[50]

Wheeler's most sensational adventure during his long Ranger career occurred in June 1907. Stationed in Yuma, he received word from Capt. Harry Wheeler to pursue suspected rustlers Lee Bentley and James Kerrick. Kerrick was a longtime badman who had served a prison term for killing a sheepherder in California. After his release he went to Arizona's lower Gila Valley, where he rustled cattle until he was sent to Yuma Territorial Prison for two years. When free again he returned to Gila country to resume his rustling activities, acquiring as an accomplice Lee Bentley, a Texan in his mid-twenties.[51]

Complaints about Kerrick and Bentley brought orders to Frank Wheeler to apprehend them. The Ranger took an eastbound train from Yuma, stopping in Wellton to wire Yuma County Deputy Sheriff Johnny Cameron to meet him with saddle horses at Sentinel. From Sentinel the two lawmen rode south toward Ajo, where Kerrick and Bentley had been stealing stock.

Wheeler and Cameron began their trek on Wednesday, June 26, riding through a barren, hostile desert in blazing heat. They found ranchers in the hills who stated that Kerrick and Bentley, calling themselves prospectors, had been spotted in the area driving stolen animals. The two lawmen rode all the way to the Mexican border in search of their prey, then turned back to the north, covering 140 miles. Saturday was the worst day — thirty-five miles of blazing heat through cacti and blistering sands. "Our horses went without water the entire day," reported Wheeler, "and the water in our canteens was so hot we couldn't drink it." [52]

By now they had picked up two Papago guides; the Indians were eyewitnesses when the lawmen jumped the outlaw camp. Knowing that Kerrick and Bentley were camped about three miles away at Sheep Dung Tanks, west of the little mining settlement of Ajo, the two officers took up the trail again at dawn on Sunday. They rode about a mile and a half then dismounted, removed their shoes, and quietly walked the remaining distance.

At the campsite six horses were staked out while the two rustlers slept, rifles close by their sides. The officers readied their rifles, then Wheeler called out a command to surrender in the name of the law. Both rustlers scrambled up, groping for their rifles. Wheeler and Cameron shouted at them to give up, but Bentley raised his weapon and triggered a shot. For a moment the desert quiet was broken by the flat explosions of rifles as each man brought his gun into play. Kerrick fired one shot at Cameron, but the deputy dropped his antagonist with the first round from his .30-.30.

Wheeler had thrown down on Bentley. When the young outlaw fired his Winchester, Wheeler cut loose, methodically emptying his magazine into Bentley. Wheeler's first slug punched into Bentley's belly, but the outlaw held his kneeling position. The Ranger pumped three more .30-.40 Winchester bullets into Bentley's torso. Somehow the stricken rustler stayed up, gamely trying to get his gun back into action. But Wheeler's final shot drilled into Bentley's left temple, ripping through his head and out his right ear. Bentley dropped face forward, dead when he hit the ground. Wheeler later testified that Bentley "showed more nerve under fire than he had ever seen displayed by a man before." [53]

Cameron and Wheeler cautiously walked over to the fallen rustlers, but both were dead. Several new Winchesters were found in the camp. The officers threw the two bodies across a pair of stolen horses, packed everything else that needed to be hauled out, and headed north. By the time they reached Ten Miles Well, a journey of twenty-five miles, the corpses had swollen badly in the heat. The officers sent word to Sentinel to wire for the Pima County coroner, but the man refused to come. The justice of the peace at Silverbell, who had jurisdiction over the Ajo area, also refused to come. While waiting for Sheriff Nabor Pacheco to arrive, Wheeler and Cameron fashioned two rudimentary coffins and lowered the bodies into temporary graves. However, the sheriff did not arrive until Monday afternoon, and even though Pacheo brought ice, by then the bodies were decomposed beyond recognition.

It was learned that both Kerrick and Bentley were wanted in Maricopa County for stealing Indian ponies. Wheeler and Cameron rode on to Gila Bend, then returned to their homes by rail. The inquest was delayed until August. Different versions of the shooting held that Wheeler had killed both outlaws or that Cameron had shot Bentley to open the gunplay. It also was rumored that the two outlaws had planned to rob a bullion shipment scheduled to pass nearby, but some witnesses stated that Bentley and Kerrick had not even stolen the six horses, paying Indians ten dollars to hire the animals. Finally, however, the coroner's verdict exonerated the officers as having "acted in the discharge of their duties and to have been justified and acting in self-defense at the time of the killing near Ajo of Lee Bentley and Jim Kerrick." [54]

There was another Ranger penetration of Papago country in 1907. On August 30 in the Papago village of El Cubo, Lariano Alvarez was murdered by an Indian named John Johns. El Cubo was surrounded by a forest of organ pipe cacti and situated beside a large waterhole in the high country twenty-five miles south of Ajo. Mescal frequently was smuggled into the isolated village from Mexico, and the men in El Cubo and other nearby communities spent much time in a drunken state. Alvarez visited El Cubo and encountered a drunken John Johns, who pulled him off his horse and stabbed him several times. Alvarez died five days later.

Tom Childs, Alvarez's brother-in-law who lived in Ten Miles Well, wrote Sheriff Nabor Pacheco about the murder. Pacheco sent an arrest warrant to Childs, deputized him by mail, and instructed him to bring John Johns to Tucson. Childs and several friends rode to El Cubo, but the Indians throughout the area were murderously hostile to the intrusion.

*A dismounted Ranger patrol late in 1907. At left is Capt. Harry Wheeler. Next to him is Rye Miles, and standing fourth from left is Oscar McAda. Both Miles and McAda were 1907 recruits.*

Childs prudently led his men out of El Cubo without trying to seize Johns and returned the warrant to Pacheco with the warning that "the Indians are up in arms, and threaten to kill the first white man that attempts to travel the Cubo trail." [55]

Sheriff Pacheco determined to exercise his authority in this remote part of his county. Pacheco frequently had worked with Rangers, and in this situation he contacted Capt. Harry Wheeler. There were at least 2,000 Indians in the area, and the two officers agreed that an expedition in force would be necessary. Wheeler quickly assembled nine Rangers: Sergeants Jeff Kidder, Billy Speed, and Rye Miles, and Privates W. F. Bates, J. A. Fraser, Travis Poole, John Rhodes, James Smith, and Tip Stanford. Wheeler himself would lead the large Ranger contingent, which meant that nearly half of the company (only twenty-two men were currently on duty) would be committed to a two-week expedition. From Pacheco's office Wheeler wrote the governor about the situation, telling him where the remaining twelve Rangers were stationed and that Lieutenant Old would command in his absence from Prescott.[56]

Sheriff Pacheco brought along one deputy, as well as a wagon and team. The twelve-man posse, armed to the teeth, rode out of Tucson at 5:00 on the morning of Sunday, September 15. On Thursday morning,

*Wheeler leading his patrol in the field.*

*Wheeler and the patrol head toward a settlement.*

their fifth day of travel, they encountered an old Indian about eighteen miles from their remote destination. Sheriff Pacheco asked him to guide the party to Gunsight, a mountain mining camp a few miles beyond El Cubo. The Indian agreed, stating that the best trail would lead through a *ranchería* called El Cubo. Pacheco replied his assent with a straight face, and about 1:30 in the afternoon the posse rode into El Cubo, deserted except for several women and children. Pacheco then innocently asked the guide to find a man named John Johns, but the Indian suddenly realized the purpose of the posse and refused to look for the murderer.

Within a few minutes several Papago men returned to the village. They were arrested and questioned, and an old man revealed that Johns was in a nearby field. Pacheco and Wheeler rode out, and as they neared the field they saw a man hurrying away through the weeds. Wheeler galloped after him, quickly seized his prey, and brought him back. The captive admitted that he was John Johns, but he accused another Indian, named Citiano, of the murder. Both Johns and Citiano were placed in the wagon and shackled into leg irons.

Many of the Indians were armed with rifles, but the twelve-man posse proved sufficiently intimidating to avert any resistance by the villagers. Six more Indians were rounded up as witnesses, and the party headed toward Tucson. Johns eventually was convicted of murder in the second degree and, in February 1908, sentenced to ten years in the territorial prison.

# 1908:
# Loss of
# a Fighter

*"I fell and was dazed, but knew that my only chance was to fight while I had cartridges left."* — Sgt. Jeff Kidder

Early in 1908 the Rangers concentrated on stopping arms sales to Yaqui Indians in Mexico. Responding to this longstanding problem, Sgt. Jeff Kidder learned of guns being smuggled from Arizona to Cananea, and Pvt. Rudolph Gunner was sent to Tucson to track down merchants who sold weapons for use in Mexico. Indeed, Ranger efforts against gunrunning continued throughout the year.[1]

While Private Gunner was in Tucson, he was asked to try to locate an illegal fan-tan game being run by Chinese men in a store on Pearl Street. Gunner twice traveled to Tucson in January to investigate the betting game, but a loud and well-trained bulldog sounded the alarm at the approach of strangers, and the gamblers could not be caught. In Prescott, Lieutenant Old located the apparent site of Chinese gambling. Old collected a few other officers, "burst in the door and found Chinamen playing something with dominoes." Frustrated, they were unable to take legal action because the game "is unknown to white men." [2]

For the past year, Wheeler had been aggravated by an underground roulette game in Naco. About once a week, usually when most of the Rangers were out of town, the roulette table would be set up late at night down an alley adjacent to the Caw Ranch Saloon, which earned at least $100 per month in profits from this irregular game. E. P. Ells, the local constable and a Cochise County deputy sheriff, provided cover for the game, presumably for a payoff. Wheeler was especially irritated that this game was carried on in the town that housed Ranger headquarters.

At last Wheeler developed an elaborate scheme to close the game. On Thursday, February 6, the headquarters detachment — Wheeler and three other men — rode out of Naco on one of Wheeler's frequent scouts. After dark, however, Wheeler and his men, reinforced by two Bisbee policemen, sneaked back into town and concealed themselves at an outhouse overlooking the presumed site of the game. There were "several fast women" present; Wheeler had paid them forty dollars to drive down from Bisbee in an automobile and encourage the roulette gamesters to open play. Wheeler crept close enough to hear the girls ask the men to "roll the ball," but for some reason they would not start a game. The lawmen waited all night, but they could not catch the gamblers in the act of playing. When Wheeler disgustedly wrote the governor about his unsuccessful scheme, he had not slept in forty-eight hours. He concluded the report by stating his determination to stay on the case: "Public sentiment here apparently is in favor of these people, but we are not influenced by the sentiment at all." A few days later Wheeler directly confronted the proprietors of the saloon, informing them that he knew the roulette game was being conducted and virtually daring them to continue their illegal activities.[3]

Wheeler was as conscientious as ever of administrative responsibilities, but he continued to ride into the field in search of rustlers and fugitives. A January scout had to be postponed when a Bisbee police officer was slain and Wheeler assisted with the investigation. A March scout would have been delayed when each of the Rangers assigned to headquarters had to be sent to various trouble spots, but Wheeler simply rode into the field alone. Another solitary mission was performed by Pvt. Sam Hayhurst, who left early in the year for West Texas in search of George Mabry, who had committed a brutal murder in Douglas in 1905.[4]

By 1908 there was widespread corruption in Yuma. District Attorney Peter Robertson urgently requested Ranger assistance. Frank Wheeler was stationed in Yuma, but it was felt that someone was needed who was unknown to the lawless element. Capt. Harry Wheeler had to travel to California on official business late in January, and on his return he stopped in Yuma. He met with Frank Wheeler, who had just caught a Tucson diamond thief and recovered $2,000 worth of jewelry. Then he conferred with District Attorney Robertson, who convinced the head Ranger of the need for undercover help.[5]

Wheeler dispatched Rudolph Gunner and R. D. Horne, who both had enlisted late in 1907. Despite their brief service they had impressed Captain Wheeler. "I don't know what I would do without these two men," said Wheeler in a January letter to the governor. "I have worked them day and night, yet they are always willing, agreeable and full of ambition, and absolutely obedient. I want to secure more like them." [6]

Gunner went out first, posing as a representative of *Everybody's Mag-*

*azine.* Robertson had been so careful of the Rangers' hidden identity that when he advanced Gunner $45 for expenses, he asked Wheeler not to mention his name on the reimbursement voucher "because I don't want him 'tipped' off." Gunner returned to Naco late in February, having collected considerable evidence against saloonkeepers who had been violating the laws against gambling and women. He reported to Wheeler that he had been in Yuma dives that he would not have left alive had his identity been known. Wheeler sent him back to continue the investigation, accompanied this time by Horne, who had just returned from official duties in Cananea.[7]

Finally, Gunner revealed his identity. He and Horne were given an office, and, assisted by Frank Wheeler, began issuing warrants to men and women who had flaunted the saloon laws. More than one citizen was angry at having been taken in by an undercover Ranger, and Gunner wisely ignored the request of an indicted saloonkeeper named Dutton who wanted him to come to the dive for a talk. "The feeling in town against us is very strong," wrote Gunner to Captain Wheeler, "more so against me than H[orne]." Robertson offered his thanks to Wheeler for sending "such good men."[8]

A few days later, however, Gunner despondently reported that the "cases came up to-day and *we* were properly Kangarooed." Both the sheriff and presiding judge acted against Robertson and the Rangers. "The Dist. Atty dont know a dam thing!" jeered the sheriff, and the judge slapped the defendants on the wrists after ruling "that the evidence is not sufficient to convict." Gunner even felt that Frank Wheeler's performance had been half-hearted, because he was hoping to be elected sheriff.[9]

Saloonkeeper Dutton bragged around town "that he was going to get that little kid," a reference to Horne, who was slight of stature. But later in the day Gunner and Horne encountered Dutton with another man in front of the Rangers' hotel.

"Say, Dutton," challenged Horne, "I understand you intend to beat me up. I'll bet if you do, you'll never do it on the Main Street."[10]

Dutton and his friend visibly became "very uncomfortable." At this point Frank Wheeler appeared on the scene, and Dutton hastily "squared matters." That evening a newspaper reporter stopped Gunner, "shook hands with me and said that he was sorry we lost out in the 'first inning,' but that the good element was with us. — I am afraid the good element is in the minority."[11]

The next day Gunner spoke with Tom Rynning, who now was warden at the Yuma prison. The former Ranger captain told Gunner that they probably would not get a jury to convict, and in succeeding days Rynning sadly saw his prediction come true. P. J. Sullivan, deputy U.S. customs collector at Yuma, drunkenly abused a prosecution witness in a saloon, attacked the man, and abruptly was walloped to the floor. District Attorney Robertson told Captain Wheeler that his life had been threatened: "I think the condition here now is about the worst I ever saw." Robertson asked Wheeler and the attorney general of Arizona to come to Yuma, and requested that Gunner be left at his side for a few weeks longer. Gunner seethed under the insults of the prosecuting attorney, while growling to his captain: "The riffraff is rejoicing and we have to stand a good deal." [12]

Mail service was excellent early in the century, thanks to a multitude of trains which delivered letters across the territory by the next day. The day after Robertson and Gunner wrote the above letters to Wheeler, the captain telegraphed that Gunner could stay in Yuma as long as needed. Gunner reported that Robertson "was jubilant; he says as long as that man stays with me I'll fight them to the bitter end." [13] Wheeler, incensed, sent a bitter letter to the governor and enclosed reports from Gunner and Robertson. Wheeler left for Yuma on the next train, pleading with the governor: "If you order me to do it, in 20 minutes I'll take possession of that town and in a day or 2 at most I'll establish law and order." [14]

But nine days later, on March 16, Wheeler returned to headquarters. [15] He had not been permitted to establish Ranger law in Yuma, and trial results had been predictably disappointing. Wheeler decided to find relief in the field. He planned a far-ranging scout, and by the end of the month he was riding through the mountains and desert at the head of a Ranger patrol.

Jeff Kidder spent most of his Ranger service in Nogales. He constantly rode the border country, hounding smugglers, rustlers, gunrunners, fugitives, badmen of every stripe. Well liked in Nogales, when he was occasionally transferred elsewhere for temporary duty the local newspaper lamented: "Jeff's many friends here are sorry to see him leave this point. He is a first-class, level-headed officer and a gentleman." [16] Detractors regarded him as an arrogant man who backed up bluster with uncommon expertise as a pistoleer. Obviously, however, many who knew him best regarded Kidder with respect and affection.

By 1908 he had advanced to first sergeant. In February he returned to Nogales from a few days in the hills, and the newspaper welcomed him back: "Jeff is a mighty efficient, experienced officer; where he is, there reigns peace and quiet." [17] By March the Nogales peace and quiet was threatened by "a number of undesirables," but Sergeant Kidder simply ran them out of town. [18]

Soon Kidder was again in the field, returning to Nogales on Wednesday, April 1. He stabled his horse at Al Peck's Livery and walked up Grand Avenue to the post office for his mail. A letter from Captain Wheeler, currently leading a sweep for outlaws through the Chiricahua Mountains, reminded Kidder that his enlistment expired each April 1, and the veteran Ranger was invited to meet Wheeler at headquarters in Naco to be sworn-in for another year. Kidder, always rough with lawbreakers, was under "some minor charges preferred by the criminal element," but there had been complaints against him before and he had every intention of remaining a Ranger. [19]

*Ranger Jeff Kidder holding the reins of his horse. His pearl-handled, silver-plated .45 is resting on his hip.*

On Thursday Kidder scooped up little Jip and placed his ever-present mutt across his saddle. Jeff rode east out of Nogales with a pack animal in tow and headed across the Patagonia Mountains. He made it to John Sutherland's Bootjack Ranch, where he spent the night with the old bachelor. Jeff finished his journey on Friday, April 3, riding into Naco during the afternoon. He checked in at headquarters, spent some time there with Sgt. Tip Stanford, then announced that he was going to cross into Mexican Naco to meet a friend coming up from Cananea. Stanford surmised that the "friend" was one of Kidder's numerous informants.[20]

There had been malignant resentment against *gringos* in Naco, Sonora, since the 1906 troubles in Cananea. Wheeler told his men they would most likely be killed over there and ordered them not to cross the line. Kidder, intending to go anyway, wanted to appear inoffensive while still keeping a means of self-defense. He stripped off his gun rig, concealing the silver-plated .45 in his waistband beneath his coat, and punched six cartridges out of his ammunition belt into a pocket. Then he headed on foot toward the border entrance. Kidder crossed into Sonora accompanied by friends and by Jip, bouncing along at his master's heels.[21]

Kidder may or may not have encountered his friend, but he made a valiant effort to find him. He investigated several *cantinas*, visited various bartenders, and fandangoed with a number of dance hall girls. By midnight he was in a *cantina* where an American had been shot three months earlier. Kidder retreated to a back room with "Chia," a new girl in Naco. (Later he stated that he was asking Chia where to locate a fugitive.) After a time Kidder was ready to leave, but when he checked his vest pocket he found that his last silver dollar was missing. With a certain lack of chivalry Jeff accused Chia of stealing his dollar. Infuriated, she hit Kidder, screamed for the *policía*, and jerked open the door. Immediately two officers, Tomas Amador and Dolores Quias, burst into the room, revolvers in their fists.

Amador squeezed off a round and the bullet punched into Kidder's middle, entering just to the left of his navel and ripping out his back. The impact of the heavy slug hurled Kidder off his feet, but even as he crumpled to the floor he palmed his Colt. Sitting on the floor, Jeff opened fire and dropped both policemen with bullets in the knee and thigh. Neither officer had any fight left, and Jeff crawled to the door and struggled to his feet, followed by Jip.

Stunned and bleeding, Jeff stumbled outside and tottered shakily toward the border, a quarter of a mile away. Several Mexican line riders blocked his way, triggering Winchester bullets at him. He cocked his .45 and pulled the trigger, but the gun was empty. Under heavy fire, Jeff cut to his left and made the fence.

Too weak to struggle through the wire, he reloaded his pistol with his last six cartridges. Victoriano Amador, chief of police and brother of the officer who had drilled Kidder, headed the attack. Jeff nicked him in the side, then held the others at bay until he ran out of ammunition. Finally, he shouted out that he was "all in."

Chief Amador led nearly a score of men to Jeff, who stood unsteadily until one of his adversaries clubbed him with a revolver. The wounded Kidder toppled to the ground, then was dragged about fifty yards away from the fence and again was pistol-whipped. One angry officer wanted to blow Kidder's brains out, but the wounded police chief halted this final brutality. At last Kidder was dragged all the way to the *juzgado local* and dumped into a cell. The vindictive officers rifled Kidder's pockets, taking his badge, a pocket watch he had received for graduation from high school, his Masonic key and, of course, his fancy Colt. Kidder was offered no medical attention and was not even given a blanket. Gutshot and bleeding, he spent the next two hours alone in a cold cell.

When word of the fight reached the American side, Mexican officials were quickly contacted. A judge named Garcia permitted the removal of Kidder to a private home, although Jeff was not released from Mexican custody. A Mexican officer armed with a Winchester was stationed near his cot as a doctor tended Kidder's wound. At dawn on Saturday, another physician from Bisbee crossed the line to aid Kidder. Their diagnosis was grim: Kidder's intestines apparently had been perforated and the most optimistic opinion they would venture was that he might recover.

Throughout the day Kidder was visited by friends and fellow officers

*The Naco, Sonora,* (cantina) *where Jeff Kidder shot it out with two Mexican policemen.*

(Courtesy of John Payne, Bisbee, Arizona)

*By the mid-twentieth century the building where Kidder was shot had been renovated and converted into a grocery store. Today the old structure stands vacant and dilapidated.*

(Courtesy of John Payne, Bisbee, Arizona)

as Tip Stanford telegraphed several points in a vain effort to contact Captain Wheeler. A reporter from the Bisbee *Review* interviewed Kidder, who talked freely with the journalist and his other visitors while his dog faithfully lay beside the cot.

"I know that a great many people think I am quick-tempered and without looking into the details will form the opinion that I precipitated this trouble. It is probable that I may die, and I would like the public to hear my side of the affair."

While relating the story of the shooting and of his treatment after surrendering, the wounded Kidder commented bitterly, "If anybody had told me that one human being could be as brutal to another as they were to me I would not have believed it.

"It's too bad such an unfortunate thing occurred," concluded Kidder, "but if I am fatally wounded, I can die with the knowledge that I did my best in a hard situation."

Deputy U.S. Marshal John Foster, who had been the Ranger lieutenant when Kidder enlisted five years earlier, was by Jeff's side. "You know Jack," said Kidder, "that I would have no object in telling what is untrue. They got me, but if my ammunition had not given out, I might have served them the same way." [22]

Kidder seemed to hold his own throughout Saturday, but during the night he weakened noticeably. He repeatedly left word for Wheeler that he had disobeyed orders by going into Mexican Naco only because he was searching for a fugitive. A little after 6:00 Sunday morning, thirty hours after he was shot, Jeff Kidder died at the age of thirty-three.

Mexican authorities at first refused to release Kidder's body. The shooting aroused violent feelings on both sides of the line, and reportedly as many as 1,000 Americans were ready to march across the border to retrieve Kidder's remains. American fury was fanned by rumors that Kidder had been deliberately lured into a fatal trap, perhaps because of his overbearing manner or perhaps because Mexican officers were jealous of his successful actions against border lawbreakers.[23] Officials were trying hard to keep a lid on the explosive situation, but the right leadership would have triggered an armed invasion. Wisely, Judge Garcia granted permission to release Kidder, and the body was taken to the Palace Funeral Parlor in Bisbee.[24]

The dead man's skull was so badly battered that embalming fluid ran from his eyes and nose. "Kidder's head must have been almost jelly," announced the undertaker. One three-inch laceration had gone to the bone in his forehead, and another swollen wound was clotted with blood on the right side of his neck. Several ribs had been broken by kicks, while the backs of his hands were torn, apparently from trying to ward off blows to his head and body.[25]

Captain Wheeler still could not be located. Since March 18 he had been absent from Naco on one of his periodic sweeps. He was accompanied by Sgt. Arthur Chase and Privates Rudolph Gunner and R. D.

Horne, leaving Sergeant Stanford alone at headquarters. Wheeler and his detachment rode through Santa Cruz, Pima, Pinal, and Cochise counties, as well as southwestern New Mexico. For weeks the Rangers combed some of the wildest and most barren portions of Arizona, searching for rustlers and, on the day Jeff Kidder died, arresting a Mexican wanted for murder in Sonora. Early Tuesday morning, riding back through the broad Sulphur Spring Valley east of Tombstone, the Rangers encountered two cowboys who told them about Kidder. It was twenty-eight miles to Bisbee and both men and horses were fatigued from their long trek through rough country, but Wheeler led out immediately. Before noon the Rangers and their prisoner arrived in Bisbee on jaded mounts.[26]

Wheeler and his men went directly to the undertaking parlor. Kidder lay in an open coffin, his face plainly showing the effects of a beating despite the ministrations of the undertaker. None of the four Rangers spoke, but their expressions were bitter. Tears came to Wheeler's eyes. After the funeral services he said: "Jeff Kidder was one of the best officers who ever stepped foot in this section of the country. He did not know what fear was . . . and was hated by the criminal classes because of his unceasing activity in bringing them to justice." [27]

Sergeant Stanford had wired Kidder's mother in San Jacinto, California, on Sunday, then wrote a letter giving details. Mrs. Kidder asked that Jeff's body be shipped to California for burial. On Wednesday, April 8, 259 members of the Bisbee lodge of Elks escorted Kidder's coffin to the depot, and one of the Elks accompanied the body to San Jacinto. Jip followed his master's coffin to the depot, where Sam Hayhurst picked up the little dog with the idea that it would become the Ranger mascot. But Jip mourned and continually ran away, trying to find Kidder. Captain Wheeler passed a hat and raised more than enough money to send Jip on a train to California.[28]

Wheeler wasted no time in going to Naco, Sonora, to investigate Kidder's death. The Bisbee *Review* reporter already had interviewed the Mexican policemen and other principals through a translator. Wheeler now spoke with the three wounded officers directly, as well as with numerous eyewitnesses. The stories conflicted wildly and, of course, put Kidder in an unfavorable light. In one account, a bartender in the *cantina* stated that Jeff "had taken at least fifty drinks." [29]

It took a trip to Cananea, but Wheeler managed to recover Kidder's revolver and badge, the latter which was found in a pawn shop. Mexican authorities, trying to smooth over the explosive border situation, helped Wheeler locate Kidder's effects.[30] The saloons of Naco, Sonora — about fifteen in number — were ordered closed, and twenty policemen and line riders were dismissed, including the wounded chief. A Tucson *Citizen* headline melodramatically announced: "RANGER'S BLOOD WASHES AWAY A REIGN OF VICE." [31] One of the other wounded policemen suffered complications and died,[32] but Wheeler was irked to learn that some of the discharged officers soon were reinstated. The Ranger captain immediately

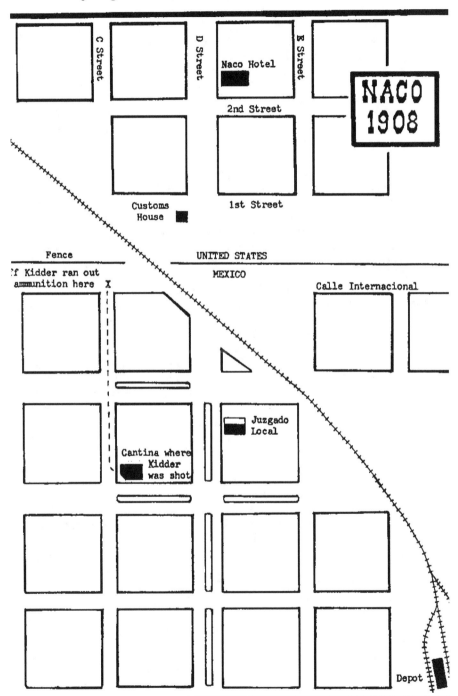

Naco Hotel

C Street
D Street
E Street

NACO
1908

2nd Street

Customs
House

1st Street

Fence

UNITED STATES

f Kidder ran out
ammunition here    X

MEXICO

Calle Internacional

Juzgado
Local

Cantina where
Kidder
was shot

Depot

requested Mexican officers to "refrain from coming over here armed"; they were to send him a note asking his assistance in legal matters. Wheeler again forbade his men to cross into Mexico, then he sent Kidder's gun and discharge papers to the family in California.[33]

Kidder's discharge was the source of controversy. Mexican authorities had inquired if Kidder was an officer in good standing on April 3, the date of the shooting, and the Ranger captain had replied affirmatively. However, Kidder's enlistment ran out three days before the gunfight. He had gone to headquarters in Naco to reenlist, but he could not complete the process until the captain returned. Ranger reenlistment procedure was to submit discharge papers to the governor for his signature after each year's service; at the same time the governor would sign an enlistment for another year, if the Ranger wanted to stay in the service. It was intended that service would be continuous, and Wheeler pointed out that in the military, if a brief interval occurred between discharge and reenlistment, customarily the soldier's service was extended until he was actually discharged. Wheeler grudgingly conceded, however: "Strictly speaking, I would feel he was not in the service . . . the day of his wounding."[34]

Technically, therefore, Jeff Kidder was not regarded as a member of the company when he was slain. Carlos Tafolla, shot during the second month of the company's existence, retained the dubious distinction of being the only man slain as an Arizona Ranger.

In May, Captain Wheeler closed in on a horse thief named George Arnett. Arnett, a thirty-seven-year-old hardcase who sometimes used the alias George Wood, had been stealing horses for several months in Cochise County and disposing of them across the border in Mexico. Suspected of robbing two Chinese men near Tombstone, Arnett also had bragged about looting a gambler in Lowell, and he had served a penitentiary term for horse theft. Wheeler had arrested him in March for robbing the Chinese men, and he considered Arnett "the worst man in Cochise [County]."[35]

Arnett blamed Mrs. Edward Payne of Lowell for revealing the robbery, and when he threatened to shoot her and her husband, she asked area lawmen for help. Acting on a tip from an Arnett accomplice, Wheeler and Deputy Sheriff George Humm, who was regarded as an expert marksman, set a trap for the suspected horse thief in a canyon about a mile east of Lowell, which is just a mile east of Bisbee. For four nights in a row Wheeler and Humm held a fruitless vigil. On the fifth night the two lawmen left Lowell about 2:00 A.M., Wednesday, May 6, and rode toward a rendezvous along different trails. Each man tied his horse in the mountains and walked to the canyon, carrying "bull's eye" lamps. When they met, Wheeler concealed himself behind a big saguaro cactus, while Humm ducked behind a bush.[36]

About an hour later, the officers heard someone approaching. A horseman topped a rise, leading a saddled mount. (Wheeler later testified his unconfirmed impression that a second thief was present, but vanished when the shooting started.) When the rider and his two animals were less than twenty feet away, Wheeler and Humm beamed their lamps at him, ordering him to throw up his hands. Instead he shouted, wheeled and spurred his horse, and snapped off a shot which clipped a bush between the lawmen. Wheeler already had leveled his revolver, and he held his lamp in his left hand at arm's length to distract the outlaw's aim. When the fugitive fired, Wheeler instantly triggered his .45, and he heard Humm's revolver go off beside him.

The retreating outlaw fired a second pistol shot from about fifty feet away. Humm scrambled for his rifle and squeezed off one round as the outlaw galloped up the mountainside and disappeared over a ridge.

The lawmen thought their prey had escaped unscathed. Their horses were secluded at some distance, so they hurried on foot back to their mounts and met in Lowell. The two officers rode back to the canyon, and as they crossed the ridge they spotted the outlaw's two horses, one of which had been wounded in the hind legs.

Realizing that the horse thief must have been injured, they carefully began to search the surrounding area. About 4:00, by the light of their lamps, the officers located Arnett's corpse no more than a quarter of a mile from the site of the shooting. He had been shot twice: one slug passed through his left arm below the elbow, while the fatal bullet ranged through his left shoulder, into the left side of his neck, and emerged from his right ear. A coroner's jury was quickly brought to the canyon, and after they viewed the remains, Arnett was hauled into town by the Palace Undertaking Company.

That afternoon the inquest was conducted at the coroner's office, and Wheeler testified: "I have heard a relative state that Arnett had said he would never submit to arrest." Wheeler and Humm, of course, were exonerated since it was "the general opinion of the public that a dangerous man has met his end." [37]

Wheeler and Humm had received inside information about Arnett's activities from a gang member known as Charles Coleman. The stoolie's real name was Charles W. Heflin, a thirty-year-old criminal from Navarro County, Texas. Heflin evidently was willing to trade outlaw secrets for legal clemency. Wheeler and Sheriff John White had contacted "Coleman" about doing undercover work, and after proving himself by helping to trap Arnett, Heflin seemed to be a candidate for further undercover work.[38]

Wheeler planned a sweep into Sonora, where Arnett and his confederates had taken several hundred horses that were stolen in Arizona. Heflin's services as a guide were imperative, and Wheeler temporarily placed the outlaw on the Ranger rolls in May. "He is not a man I would for one instant have in my Company," explained Wheeler, who commanded Heflin to sign an agreement specifying the limitations of his enlistment. Ranger Heflin was to be "shorn of all authority to act as an officer." His

Ranger papers were to read "C. W. Heflin," and his identity was to be kept a strict secret.[39]

Heflin revealed that Constable Ells of Naco, with whom Wheeler had clashed before, had been associated with Arnett, and that the two had planned a payroll robbery, aborted because of Arnett's death. Wheeler also learned that Arnett had been protected by a high Mexican official who owned a *rancho* near Moctezuma, 150 miles south of the Arizona-Sonora border. Arnett had personally delivered many stolen horses to this official, and Heflin promised that the animals could be found in the man's pastures.

Soon area badmen began to sense that the authorities were receiving inside information, and various outlaws deserted the Bisbee-Naco vicinity for Sonora. Pleased at this exodus of so many "tough characters," Wheeler commented that "Heflin is safe." [40]

Wheeler had almost completed his preparations for the expedition into Sonora when Mexican officials notified him that it would be necessary to delay his entry until June 10. Disappointed, Wheeler rode into the mountains for a few days on a scout, then returned to receive the pleas of a Mrs. Fikes, who had just learned that her stolen horses were pastured near Moctezuma. Armed with a personal letter from Gen. Rafael Torres and accompanied by Mrs. Fikes's son, Wheeler eagerly traveled to Douglas to meet a Mexican major and several enlisted men across the border. Local sportsmen, knowing that Wheeler was an expert marksman, tried to persuade him to stay over for one day to bolster the Douglas Rifle Club in a shooting contest against the Fort Huachuca team. But Wheeler, impatient of further delays, insisted upon leaving that night, Friday, May 29. "It is a wild and dangerous country down there — no law — no officers — little known and a favorite resort of renegades — I hear a dozen Mexicans down there are wearing crepe on their hats over George Arnett being killed —" The fearless captain of the Arizona Rangers found this lawless, threatening environment an irresistible challenge.[41]

Within less than a week Wheeler had traveled to Moctezuma, secured a number of horses, and started north with five stolen animals. Exhilarated, he was convinced that the thieves were frightened. Stopping at Nacozari to rest his mounts, he wrote to the governor's office that "you should see the looks I get sometimes of the tough element, which to me, is a matter of pride." [42]

Wheeler obtained twenty days' leave of absence from the Ranger service for himself and Sgt. Rudolph Gunner, as well as C. W. Heflin and George Humm. Impressed by Humm during the fight with George Arnett, Wheeler had arranged for the deputy to be placed on the Ranger rolls from June 10 to June 20, 1908. Heflin, serving as guide, was not allowed to carry firearms.[43]

Wheeler, Gunner, Humm, and Heflin rode out of Naco into Mexico on Thursday morning, June 11. They were accompanied by a Mexican army escort, including a major and a lieutenant. Averaging fifty miles a day, they plodded through sizzling summer heat across mountain trails. Rising each day at 3:30, within an hour they were in the saddle and did not pitch camp until 8:00 P.M. After nine days of hard riding, penetrating mountain hideouts, and crisscrossing rustlers' haunts, the exhausted party entered Moctezuma. Three of Arnett's rustling accomplices were captured, along with four stolen horses; the rustlers were wanted on Mexican charges, and Wheeler sent them to Hermosillo.[44]

Wheeler never really seemed to mind the hardships, and he always relished a head-on challenge with outlaws. With a glint of triumph he reported: "The Arnett gang is now broken up and the criminal element is badly frightened." [45] Harry Wheeler genuinely enjoyed intimidating badmen.

In 1897 Charles Bly, an inmate in the New Mexico Territorial Prison in Santa Fe, escaped confinement, a year into his four-year sentence, by riding off on the warden's horse. Resolving to change his ways, the fugitive assumed the name "Frank G. Sherlock" and sought a fresh start in western Arizona. By 1900 he had wangled a deputy sheriff's appointment in Mohave County, and in eight years' service he "had run down many criminals." Concurrently Sherlock also served Mohave County for four years as livestock inspector, a position to which he brought insight, since he had been sentenced to prison in New Mexico for rustling.

Deputy Sheriff Sherlock was a conscientious officer and he was praised for leading "an exemplary life," but finally his past caught up with him. Sherlock was performing some contracting work for the Grand Canyon Lime and Cement Company at Nelson. A former fellow convict worked for the company and Sherlock arranged his discharge, perhaps because he feared the man or perhaps because the ex-con deserved firing. For whatever reason, the man vindictively tipped off the Rangers that Sherlock was a prison escapee.

Lieutenant Old and Private Herb Wood, both stationed at Williams, traveled to Nelson on July 21 and found Sherlock at the cement company's camp. Knowing Sherlock to be "a dead shot," the two Rangers determined to make their arrest as safely as possible. Old, who knew Sherlock, innocently strolled up and introduced Wood. When Sherlock and Wood shook hands, Old drew and cocked his revolver in a smooth motion and jammed the muzzle against Frank's stomach.

"Hands up!" ordered Old. "I have a warrant for your arrest."

"Guess you got me, kid," replied Sherlock as he raised his arms.[46]

Wood extracted a gun and extra cartridges from Sherlock, as the ex-convict digested what had happened. When the Rangers led him away he began to weep openly. Sherlock admitted his true identity, emphasized his positive lifestyle for more than a decade, and asked the Rangers to speak a good word for him with New Mexico Governor George Curry. Soon Sherlock-Bly was in the custody of Captain Christman of the New Mexico penitentiary, heading east on a train.

But Sherlock was not long behind bars. Governor Curry investigated the well publicized case (newspapers called him "Arizona's Jean Valjean" after the harried hero of Victor Hugo's *Les Miserables*), then summoned Sherlock. On August 18 the prisoner was granted a complete pardon. Sherlock began weeping, then through his tears told the governor that he would return to Arizona to his contracting business.[47]

William Downing was a longtime Arizona badman whose belligerent nature and defiance of the law inevitably brought him into conflict with the Rangers. A Texan, Downing moved to Arizona late in the nineteenth century, radiating rumors of past misdeeds: everyone assumed that his name was an alias; it was widely alleged that a reward of $5,000 was on his head; and many people thought he was Frank Jackson, the only member of the Sam Bass Gang who had escaped the bloody shootout in Round Rock, Texas, in 1878. He brandished "scars of former battles, as one of his legs in which he had been badly shot troubled him all the time." [48]

Downing established a ranch near Willcox, and most of his neighbors suspected that his primary livestock activities were illegal. Despite his reputation he obtained appointment as a deputy constable, and with Constable Burt Alvord planned a train robbery at Cochise Station, a whistle stop ten miles southwest of Willcox. Later Alvord, Downing, Billy Stiles, and other members of the gang were arrested and convicted of the holdup. When Alvord and Stiles engineered their infamous breakout from the Tombstone jail, Downing inexplicably was one of the men who chose to stay behind bars.[49]

During the seven years he served at Yuma Territorial Prison, his wife exhausted every effort to gain his release. She sold their ranch, found work as a servant in Tucson, then died of heart failure before Downing was set free.[50] Downing returned to Willcox in October 1907 and somehow scraped up the resources to open the Free and Easy Saloon, where crime became rampant. Rumor held that Downing engineered the robbery of his customers from time to time and that he had resumed dabbling in rustling. A southpaw marksman, he packed a revolver in his left hip pocket and clashed openly with many local men, including Willcox Constable Bud Snow and Ranger Billy Speed, who was threatened on more than one occasion by the bellicose Downing.[51]

William Slaughter Speed had been with the Rangers since 1906, but neither Wheeler nor Rynning had been impressed with his performance. Both Ranger leaders put up with him because he was an expert cowman who provided excellent service to the Livestock Sanitary Board. Curiously, Captain Wheeler promoted Speed to sergeant in the hope that higher rank would improve his qualities as a Ranger. This experiment proved unsuccessful, and the thirty-seven-year-old Ranger was demoted to private.[52]

Despite his unsatisfactory two-year tenure as a Ranger, within a few months Speed proved to be one of the few men in Willcox who would not be intimidated by the reputation and threats of William Downing. Billy

*By 1908 William F. Downing was the most troublesome badman left in Arizona.*

Speed remained mindful of Captain Wheeler's admonition that "if anyone must be hurt, I do not want it to be the Ranger." [53]

The festering predicament with Downing finally came to a head in the summer of 1908. In July the local justice of the peace wired Captain Wheeler twice in one day, stating that Speed and Constable Snow were absent on duty elsewhere and that Downing was drunk and terrorizing the town. Wheeler immediately dispatched Sgt. Rudolph Gunner and Pvt. John McKittrick Redmond to Willcox, ordering them to shoot Downing at the first flicker of trouble. By the time they arrived Downing had settled down, but Gunner and Redmond stayed in town for several days until the badman promised to behave. Nevertheless, Wheeler wrote Speed with emphatic instructions "to take no chance with this man in any official dealing you may have with him." Wheeler left no doubt as to his meaning: "I hereby direct you to prepare yourself to meet this man . . . , and upon his least or slightest attempt to do you harm I want you to kill him." [54]

Downing flaunted the restriction against women in saloons; his girls caused many of the frequent brawls in the Free and Easy and were suspected of robbing patrons. Late in July a rival saloonkeeper, George McKittrick, swore out a warrant against Downing for permitting women

*Although Billy Speed was in disfavor with Captain Wheeler, the veteran Ranger showed his mettle by taking on Downing.*

to congregate and drink in the Free and Easy. Speed and Snow teamed up to serve Downing who, after paying his fifty-dollar fine, growled to an acquaintance "that if they ever came to arrest him again, it would be a fight to a finish and that he would get them or they would get [him]." Predictably, Downing also threatened McKittrick. But at Dodge City, Kansas, McKittrick had once shot and killed a black soldier after the trooper had gotten the drop on him, and he would not be cowed by Downing. McKittrick sat up nights cradling a shotgun, and during the day kept it behind his bar, both hammers at full cock.[55]

Citizens of Willcox presented a petition at Tombstone, the county seat, that Downing's license be revoked, but no one proved willing to appear publicly against the murderous ex-convict. However, two prominent residents of Willcox arranged to meet Harry Wheeler in Benson, where the trio tried to formulate plans to control Downing.[56]

Then, on the night of August 4, Downing quarreled with Cuco Leal, one of the girls who lived and worked in his saloon. Downing hit her and gouged her eyes. She fled to the saloon of George McKittrick, who put her in a hotel room and advised her to swear out a warrant against Downing. She wasted little time in obtaining a warrant, and it was delivered to Constable Snow. Snow went to Speed's house for help, but Billy told him to save it until morning.

Downing drank heavily throughout the night, at one point mounting his horse and "raving around." Early in the morning, still drinking, he encountered Cuco and told her "he would defy anyone to come and arrest him and if anyone tried to do anything with him, he would kill somebody." [57] Two days earlier Downing had looked up Snow to tell him pointedly that he wanted the Ranger to stay away from his saloon, "and if Speed ever stuck his head inside of the door he would shoot [it] off and if he does not come in, he would kill the son of a bitch, anyway, when the time came." [58]

Early the next morning Speed walked downtown for a shave at the barber shop — with his Winchester nearby — as Snow came up the street looking for the Ranger. Downing, in his shirtsleeves, was still inside the Free and Easy waiting for the lawmen to come. When he told a customer, R. E. Cushman, that he was about to be arrested, the customer "told him to act like a man and go down and pay his fine." As Downing started outside he reached behind the bar for his six-gun. "I told him not to be a dam fool," said Cushman, and Downing put down his weapon and left through the back door.[59]

Snow saw Downing emerge from the Free and Easy, and the constable ran to the barber shop to tell Speed. The two lawmen noted Downing hurrying toward a shack near the jail, about a block from his saloon. Snow looped around behind the jail, while Speed advanced up the boardwalk toward the shack. Speed turned up an alley, whereupon Downing slipped out of the shack and went into the street. A bystander shouted to Speed that Downing was coming up the street, and the Ranger headed off his prey. In the middle of the street Speed stalked Downing, Winchester at his shoulder. Speed ordered Downing to throw up his hands. Downing raised his arms and walked unsteadily toward the Ranger.

Apparently, in his liquor-befuddled state he forgot that he had left his six-gun behind. When he was less than thirty feet from Speed, Downing's arms went down. He began to grope with his left hand at his hip pocket as Speed again shouted for him to throw up his arms. But Downing doggedly advanced, feeling for the gun that was not there.

Billy Speed pressed the trigger of his Winchester. The .30-.40 slug penetrated Downing's right breast and ripped out beneath his right shoulder blade; the impact of the bullet threw him onto his back. Speed came to his side, rifle at the ready, as Snow and several bystanders ran to the fallen man. Within two or three minutes, Downing was dead.[60]

Judge Page quickly assembled a ten-man coronor's jury, which included ex-Ranger Bud Bassett. They opened Downing's shirt and examined the wound. Then Downing's pockets were carefully searched: they found a watch, a wallet, rings, money folded inside handkerchiefs, spectacles — but no revolver. It made little difference. Several eyewitnesses testified that Downing had appeared to be going for a gun, and everyone agreed that considering his repeated threats Speed would have been suicidally foolish not to have fired. The jury concluded that Speed "was perfectly justified in the act and therefore we exonerate him from all blame in the matter."[61]

Wheeler was telegraphed about the killing. He wired the governor's office, then left on the first train for Willcox. In his report he praised Speed and emphasized that everyone in town was afraid of Downing. Wheeler remarked that "this is the first time I have ever known a dead man to be without a single friend and the first time I have known a killing to meet absolute general rejoiceing [*sic*]."[62]

Other Ranger arrest attempts in 1908 proved less conclusive. William C. Parmer enlisted in the Rangers in July and was stationed at Benson. In November the twenty-five-year-old Ranger learned that a notorious thief, Will Van Valer, was hiding at the Harmon Ranch a mile and a half northwest of town. Van Valer was twenty-two, a six-foot-two, 180-pounder who had been reared in the Benson area. He had committed numerous burglaries and open robberies, and several warrants were in force against him, but he had proved elusive, escaping from various officers on five occasions. Once he had been apprehended by Ranger Oscar McAda and placed aboard a train for the county jail in Tombstone, but he managed to escape from the train.

On Sunday, November 15, Parmer rode onto the Harmon Ranch about dusk. A short distance from the ranch house Parmer encountered Van Valer. Parmer shouted at Van Valer to throw up his hands, but the outlaw defiantly drew his revolver and snapped off a shot. Parmer fired two rounds from his Winchester in return, winging Van Valer in the ankle and wrist. The fugitive dropped his gun but, slippery as ever, vanished into the mesquite growth.

Parmer followed Van Valer's trail by looking for blood stains. Darkness soon set in, however, and the Ranger returned to Benson to organize a search party. Van Valer went to his family home, but before he could even have his wounds dressed, officers showed up at the front door. He slipped out the back door and escaped town on a freight train. At the Van Valer home Parmer found considerable stolen property, which was turned over to Benson authorities.[63]

Harry Wheeler himself suffered a similar escape. In December, following a good piece of detective work, Wheeler and another officer arrested a murderer named Christo Davlandos. Davlandos was a Greek who lived in Cananea, but the arrest apparently was made in Naco. After Davlandos was handcuffed, the lawmen began to walk their prisoner toward jail. Suddenly, Davlandos broke away, and even though his hands were cuffed behind his back, he outdistanced his pursuers. "I could easily have shot him," said Wheeler, "but I couldn't do it." [64] Wheeler, who often had fought head-on duels with outlaws, refused to shoot the fleeing Davlandos in the back because he was handcuffed and because he was afraid the man might be innocent. When it came down to it, the deadly Harry Wheeler demonstrated ultimate respect for the American legal system.

During 1908 Captain Wheeler feared that he would lose key men to other law enforcement jobs. Several ex-Rangers already filled various officers' positions around the territory, and as respect for the Rangers grew there were efforts to secure the services of members of the company for numerous lawmen's offices. Wheeler wrote to the governor's secretary stating that many citizens of Santa Cruz County were trying to persuade Lieutenant Old to run for sheriff. He also thought "it extremely probable that [Frank] Wheeler will be the next Sheriff of Yuma County," and there was talk of making Tip Stanford constable of Naco, Rye Miles constable of Benson, Bob Anderson city marshal of Globe, and Sam Hayhurst constable of Douglas.[65]

Harry Wheeler himself was approached by numerous citizens, both Democrats and Republicans, to run for sheriff of Cochise County. "I have no time to run around playing the [political] game," wrote Wheeler. "I cannot pay attention to my business and work for my political interests at the same time but I know that I could win this next election." Despite his protests, Wheeler discussed at length his election possibilities.[66] Clearly, the political bug had bitten Wheeler, but he stayed with the force, and so did Billy Old and Tip Stanford and Sam Hayhurst. There was a mystique about the Ranger company that bound many of the best men to their five-pointed stars.

Personnel turnover was average during 1908. About a dozen men left the force and about the same number enlisted, including C. W. Heflin, designated as a Ranger for a brief time while he betrayed the Arnett gang

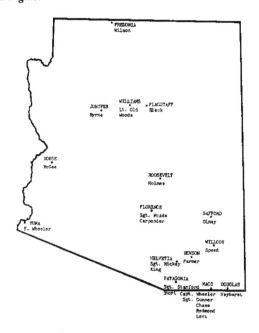

*RANGER STATIONS IN SEPTEMBER 1908. Twenty-three men were on the roster. There still was a concentration of Rangers in the southwestern part of the territory. Ranger headquarters was at Naco. The small Northern Detachment was commanded by Lieutenant Old from Williams.*

of horse thieves, and Deputy Sheriff George Humm, who signed up on a temporary basis to accompany Harry Wheeler into Mexico.

To save money and thereby allay critics of the Rangers, the approved strength of the force during the year was a maximum of twenty-three officers and men. Even though outlawry was noticeably declining in Arizona, Harry Wheeler continued to be barraged with petitions for Rangers. "As you know, Sir," wrote Wheeler to the governor in June, "the demands made upon the ranger service are constantly becoming more numerous, and I am unable to comply with many requests made upon me." [67] The surviving Ranger correspondence for 1908 is filled with pleas for Rangers. In August, Wheeler reported, "I have thirty requests for Rangers, or more, that I cannot honor, until some of the boys finish present work." [68] Wheeler himself was often in the field, and not infrequently he had to scatter his headquarters squad to various points in response to pleas for Rangers. When the Rangers were forced to abandon Naco for any length of time, invariably "some depredation" was committed by local criminals. [69]

A squad of Rangers was stationed in Williams for a time during the spring. The Rangers antagonized many people by arresting every man they found carrying a gun or knife. But at the next court session there was no murder case in Williams for the first time in twenty years, and the Rangers were praised by responsible citizens and the newspapers. Rangers were sent in June to Octave, a mining town in southwestern Yavapai County. The Rangers made a dozen arrests, many of them for fighting, and quickly calmed Octave. At the end of the year the sheriff of Cochise County requested that Rangers be sent to the new copper mining camp of Courtland. Several hundred men already had hustled to Courtland, with twenty to thirty new arrivals daily. The only jail facility was a mine tunnel, but Wheeler sent two Rangers who promptly established an atmosphere of "perfect peace and quiet." [70]

At the first of the year Pvt. Lew Mickey was stationed at the construction site of Laguna Dam, located on the Colorado River twelve miles above Yuma. He was paid fifty dollars above his monthly Ranger wages by the construction company, and the investment paid off for them. During the first six months of 1908, Mickey was not required to make a single arrest. With the roster reduced, in practice, to twenty-three men, Captain Wheeler could not spare a Ranger for such limited service, and he began maneuvering to transfer Mickey elsewhere. Frank Wheeler was stationed at Yuma, forty minutes away, and fewer than 100 persons worked at Laguna. [71]

Captain Wheeler regarded Lew Mickey and J. T. Holmes as the best candidates for the sergeants' vacancies created by the death of Jeff Kidder and the departure of Rye Miles when his enlistment ended in April. Mickey eagerly accepted elevation to sergeant, as well as a transfer from lonely Laguna Dam. But Rudolph Gunner, who had been in the company only a few months, received the other promotion when Holmes chose to remain at his home in Roosevelt rather than transfer to a sergeant's station.

In October Captain Wheeler returned to Naco from a long scout and was shocked to discover his headquarters clerk, Pvt. A. F. Chase, "under the influence of liquor, and upon the Mexican side, contrary to my strict orders." Wheeler was fond of the thirty-year-old Chase, regarding him as "worth two ordinary men" and stating that he was "my favorite Ranger, of all I have ever had." But even though the captain was grieved over the situation, he reported the incident to the governor and called for Chase's discharge.

For once, however, the rigid captain had second thoughts. Numerous citizens and fellow Rangers pleaded with Wheeler not to fire the big, likeable Chase, and it took little intercession to save him. "I fought down some personal habits years ago," wrote Wheeler to the governor. Wheeler recounted Chase's assets: "He has experience both as a soldier and as a handler of stock, is well educated and comes of an excellent family." Wheeler proposed to suspend Chase's pay for a month, "and have him do all the extra work around the quarters and Camp." Throughout the history of the Rangers, the governor had permitted the captain final say on personal matters, and Kibbey gave his permission for Chase to remain in the company.[72]

The next month Wheeler was further disappointed to learn that fondness for the bottle had caused Sgt. Rudolph Gunner to neglect his duties. Wheeler abruptly broke his straying protege to private, although Gunner reenlisted the next month.[73] November revealed another Ranger to have a drinking problem. Samuel Black, a forty-five-year-old West Virginian who had served as city marshal of Flagstaff, was enlisted by Lieutenant Old in April 1908. Although Black was a promising Ranger (Wheeler was especially impressed when Black arrested his own brother-in-law), late in November Old reported to Captain Wheeler that the man had been discovered drunk. Wheeler had not yet met the new private, who could not be reduced in rank, but the captain immediately discharged Black from the force.[74]

A. E. Ehle, who had been a Ranger private for less than nine months, left the company on May 1. He had received a cash inheritance and decided to raise oranges in California, rather than chase criminals in Arizona. Wheeler was sorry to lose Ehle, as well as veteran Bob Anderson, who tendered his resignation at the same time. Anderson had enlisted in October 1902, and only Frank Wheeler, who had signed up one month earlier, had served longer than the forty-two-year-old Tennessean. Harry Wheeler managed to talk Anderson into serving out his sixth enlistment, but in October, Bob turned in his badge.[75]

A more welcome departure was that of Pvt. J. A. Fraser. A native of San Antonio, Fraser came to the end of his enlistment in July. Wheeler liked Fraser no better than he liked most Texans: "I have had difficulty in compelling him to pay his debts, and in general I do not approve of his character." The rigid Wheeler had spoken, and the Ranger was terminated on July 19. Fraser, along with another Texan who had fallen from grace with the Rangers, W. F. Bates, tried to stir up trouble for Wheeler. They had little success.[76]

By this time Wheeler had developed a strong dislike for "Texans who have not been away from Texas very long." After firing Travis Poole in November 1907, he wrote the governor's secretary: "You will note all our unsatisfactory men are from Texas, . . . they seem to desire to be tough and without exception are hard to get along with." He made a similar complaint to the governor, and while he was having problems with Billy Speed, Wheeler blamed "his Texan disposition." Wheeler stated that his lieutenant concurred with his anti-Texan bias, but Old was from Texas himself, along with more than four of every ten Rangers. Texans always had been regarded by commanding officers as superb fighting men but terrible soldiers between battles, unruly and disobedient, and the commander of the Arizona Rangers clearly concurred.[77]

The exit of Pvt. R. D. Horne caused resentment among the Rangers. In April, after less than six months as a Ranger, Horne planned to return to the East as a member of a theatrical troupe, probably one of the wild

*Three members of the Rangers' Northern Detachment. Left to right, Lt. Billy Old, Pvt. Cy Byrne, Pvt. Sam Black.*

west shows then on tour. He evidently intended to use his Ranger experience on his billing. Learning of this plan, Harry Wheeler was predictably irked. Horne had served occasionally at the headquarters desk in Naco, and Wheeler prominently marked on his discharge papers that he had been a clerk. Wheeler commented to the governor: "some of the romance of the Discharge, being thus eliminated, he will be less likely to make improper use of it, by sticking it up at some stage entrance." The other Rangers were equally insulted at the cheapening of their commissions by the theatrical designs of Horne. "His comrades are all offended with him," wrote Wheeler, "as they consider their Discharge a Sacred document." [78]

In 1907 Harry Wheeler had requested a machine gun for the Ranger company, and in 1908 he asked Governor Kibbey to replace the Ranger issue 1895 Winchester with the army's weapon, the 1903 Springfield. Wheeler considered the Springfield "far above any arm," and the Rangers could obtain ammunition at half price from military posts. Wheeler also asked for a pair of bloodhounds, regarding them as a necessity in tracking down fugitives in the wilds of Arizona. Trained bloodhounds of the type Wheeler wanted could be purchased in Texas for $300, and the Ranger captain pointed out that they could be shipped rapidly by train wherever they were needed in the territory. [79]

Another request resulted from the fact that many Rangers were stationed so far from headquarters that Wheeler never got to visit with them and consult with the citizens about area conditions and their relations with the Rangers. But considering the press of his other duties, territory-wide inspection trips could be made only by train, and funds were unavailable for such journeys. "This is really my duty but it would cost me more than I would receive in an entire Months salary." [80] Wheeler asked Governor Kibbey to authorize funding for inspection trips, but opposition to Ranger expenditures was accelerating. The governor reduced the payroll by keeping the authorized roster below strength, and he did not approve expenditures for inspection trips or bloodhounds or Springfield rifles or a machine gun.

The rejections did not discourage the captain, and he continued to enjoy his work. Harry Wheeler loved to ride the open country in search of outlaws. During 1908 he ranged far and wide, repeatedly penetrating Arizona's wild mountain and desert country. He rode on long scouts and short ones; he rode at the head of Ranger detachments and he rode alone; he headed into Mexico at least three times, and he squeezed in a trip to Mexico on official business. He spent the first twenty-five days of October on a 600-mile sweep through Cochise, Santa Cruz, and Graham counties, and after a few days to rest the horses, he planned another patrol to last most of November.[81]

Like modern policemen who cruise in their black and whites through troubled neighborhoods, hoping to deter crime by their presence, Wheeler and his men observed "that hardly any crime ever occurs in the sections over which we ride and patrol." [82] On the long October patrol, Wheeler and his men caught only one malefactor, a horse thief who escaped jail a few days after the Rangers turned him over to local authorities.[83] Outlaws plainly were intimidated by the Rangers, but there was mounting evidence that the company had put a permanent crimp in criminal activities in the territory. After Billy Speed killed William Downing in Willcox, Wheeler informed the governor that Downing was "the last of the professional bad men in this section." [84]

Wheeler's report for the month of August revealed that the Rangers had made less than two dozen arrests: "The whole country seems remarkably quiet and scarcely any crimes are being committed anywhere." [85] And following the death of George Arnett and the breakup of his horse-stealing ring, Wheeler reported with obvious disappointment that "there has been absolutely no trouble of any kind and I am getting tired of so much goodness as are all the men." [86] Throughout Arizona the Rangers rapidly were working themselves out of a job.

# 1909:
# Political
# Demise

*"It would have been a pretty 'how'dy'do' if I had been killed or had killed one of them after my commission had been taken from me."*
—Pvt. John McK. Redmond

The year began badly for Capt. Harry Wheeler. On the first day of 1909 he wrote to Governor Kibbey asking for a leave of absence so that he could go to Florida. The previous spring he had made a similar request because of the desperate illness of his parents: both were afflicted with cancer, and his father had gone blind. Kibbey granted Wheeler's 1908 petition to go to Florida, but the Ranger captain had second thoughts. The older Wheelers had little money, and Harry reasoned that since a trip to Florida would cost at least $300, he would forego the journey and apply the travel funds to their needs. Besides, his brother was in Florida with their parents. Wheeler did not go to Florida even when his father died at Thanksgiving, but when his mother reached the verge of death he made the second request for leave, which was granted.[1]

Lieutenant Old, stationed now at Winslow, would command the Rangers in Wheeler's absence. Sgt. Tip Stanford would take over at headquarters in Naco, a familiar role for him because Wheeler was so often in the field. Before he left, Wheeler finalized the discharge of Luke Short, who had resigned after merely a few months' service to become a mine foreman at Sylvanite, and of Rudolph Gunner. Gunner, an intelligent and experienced peace officer, had so impressed Wheeler that he won promotion to sergeant only a few months after his enlistment in December 1907. But when Gunner's drinking problem worsened, a disappointed Wheeler demanded Gunner's resignation. "I cannot keep a drinking Ranger," stated Wheeler, pointing out that "a man who carries a gun and the Authority of the Territory must of necessity remain sober." [2] At the same time Wheeler enlisted O. F. Hicks, a thirty-six-year-old married man who had been a peace officer since he was twenty.

Wheeler finished up last-minute business on Sunday, January 3, and prepared to board a train that night. Before departing, however, he and a few other officers arrested a highwayman who had killed a woman in Cananea and fled to Arizona. Wheeler extracted a confession from the murderer, sent him across the line to waiting Mexican officers, typed out a brief report to the governor, and still had half an hour to catch his train. [3]

Captain Wheeler was gone for three weeks. Following his mother's funeral, he returned to headquarters on Monday, January 25. His most immediate problem involved two troublesome privates, E. S. McGee and Orrie King. McGee, a two-year veteran from Texas, was stationed in Bouse in Yuma County, but since his marriage he had become uninterested in Ranger duties. Just before leaving for Florida, Wheeler had sent him an order which he ignored. Wheeler was so furious that he would not even permit McGee to resign, instead insisting that he receive a dishonorable discharge from the company. The next day, January 27, Wheeler disposed of Texan Orrie King in the same way. King, who had enlisted in August 1908, came highly recommended by citizens of Douglas, but Wheeler learned unforgivable facts about his background. Immediately, Wheeler turned in a dishonorable discharge for King. King would prove to be the last man dismissed from the force by Harry Wheeler. [4]

A far more threatening problem than personal foibles had materialized during Wheeler's absence: the most serious anti-Ranger effort was in progress. He returned from Florida to find that supportive letters from seven county sheriffs and a number of district attorneys had reached his headquarters desk. On January 26 Wheeler wrote the governor's office, describing these letters and asking to whom they might be forwarded for the greatest benefit. [5]

Wheeler was able to do nothing further about the political hazard to his company. The governor's office wired him on January 27 that serious trouble was about to erupt in Globe: "Go there at once with available men." The message reached Wheeler at 10:30 P.M., and immediately he began to wire and telephone his men at their various stations. The captain and his headquarters squad boarded the first northbound train. There was no direct rail route, however, and Wheeler did not alight in Globe until thirty-six hours after his departure. [6]

By that time several Rangers already had arrived. Although in January the mountain weather was frigid, the Rangers ignored the rain and

*Early in 1909 fourteen Rangers converged on Globe to smother labor troubles in the mining town.*

biting cold, forded swollen streams, and struggled toward Globe. Sgt. Lew Mickey rode more than a hundred miles, "crossed a torrent," and arrived within less than twenty-four hours. Several other men rode horseback eighty or ninety miles in a day's time to get to Globe, and by the end of the week fourteen well-armed men wearing silver stars were patrolling the streets of the mining town.

On the night of January 28 Captain Wheeler telegraphed the governor's office that all was quiet. Unhappy miners who had been threatening disturbances were talking with mine owners and work was expected to resume in the morning. On the next day, Friday, Wheeler sent another telegram with the information that the miners indeed were returning to work, and the atmosphere was so "quiet and promising" that he was beginning to release his men to return to their duty stations. Wheeler and a few of his men stayed over a couple of days, but the crisis clearly had passed, and on Sunday all Rangers left town. The governor's secretary stated that he had felt from the beginning that the threat of trouble was "somewhat exaggerated," but the presence of fourteen Arizona Rangers doubtless had an immediate quieting effect upon disgruntled parties in Globe.[7]

On Monday, after his return to headquarters, Captain Wheeler enlisted Tom Gadberry. Despite Wheeler's stated dislike for recent residents of the Lone Star State, he nevertheless signed onto the roster a number of native Texans. Gadberry was a forty-year-old Texan who was destined to wear his Ranger badge a total of fifteen days. He was the last man enlisted as an Arizona Ranger.[8]

In 1907, when the Twenty-fourth Legislative Assembly met in biennial session, two anti-Ranger bills had been introduced. One bill, severely reducing the Ranger company, passed the legislature but was vetoed by Governor Kibbey. As the next regular legislative session approached in January 1909, it was a foregone conclusion that a determined attempt would be mustered to abolish the Rangers.

The Ranger question became a political punching bag. Since its creation by a Republican administration in 1901, the Ranger company had always been regarded by Democrats as a Republican instrument. The citizens of Arizona vigorously supported party politics during the early 1900s, as a reading of editorial pages of the period plainly reveals. Republican Governor Joseph Kibbey had aroused strong Democratic opposition throughout his four-year tenure, and by 1909 Democratic politicians recognized the abolishment of the Rangers as an obvious partisan ploy against the governor. Ominously, in the two-house Arizona legislature, Democrats outnumbered Republicans in the Council ten to two, and in the House of Representatives eighteen to six.

Campaigning for legislative seats had focused upon the Ranger question in many counties, and was especially intense in Maricopa County.[9] Elected to the Council from Maricopa County was Democrat Eugene Brady O'Neill, a bitter enemy of Governor Kibbey, and the House delegation was composed of one Republican and three Democrats, including Sam F. Webb, who was elected Speaker. Soon after the November elections, several Democratic legislators, including O'Neill and Thomas Weedin, met in Phoenix "to map out plans." [10] One of their stated aims was to do away with the Ranger company. A few days later Captain Wheeler was interviewed by a newspaper reporter and asked about the anti-Ranger movement. "I have heard very little objection to maintaining the ranger force," he said hopefully, "but it comes up in the legislature at regular intervals and has some following from certain sources." [11]

The sixty-day session was scheduled to open on Monday, January 18. By Sunday the legislators had assembled in Phoenix and met informally "to discuss proposed measures." [12] A Democratic caucus was organized, pledging, among other things, to repeal the 1901 bill which had created the Rangers. On Monday the legislature filled the speakerships, approved various appointments, and attended to other introductory business.[13]

Both houses convened on Tuesday afternoon to hear Governor Kibbey's long and carefully prepared message to the legislature.[14] Fully aware of the anti-Ranger movement, Kibbey attempted to mollify arguments against the company. He emphasized that "the Arizona Rangers have

proved so often their usefulness that it seems impossible to recommend the
repeal of the law authorizing the force." Indirectly addressing the jealousy
of local law officers, he stressed that Rangers patrolled the "many remote
sections in which the county peace officers do not ordinarily travel," but
which, of course, were haunts of lawbreakers. Knowing that Captain
Wheeler had made a specific effort to avoid conflict with county lawmen,
the governor pointed out "that the Rangers to a large extent perform functions which can not well be performed at all by sheriffs or their deputies."
Kibbey spoke of the high quality of the enlisted men, and he detailed the
number of arrests and mileage traveled during the past two fiscal years, as
well as expenses incurred. In the interest of economy, he also stated that
"it is my purpose to keep the number of enlisted men as low as possible."
(At that point Captain Wheeler had only nineteen men under his command.)

By the end of the week, however, Governor Kibbey was on an east-
bound train for Washington, D.C., to push for Arizona statehood in Congress. On Saturday, January 23, Thomas Weedin, a Democratic councilman from Pinal County who edited the Florence *Blade,* introduced bills to
abolish the Rangers and the office of public examiner. As part of the anti-
Ranger package, a bill was proposed — and eventually passed — creating
"Deputy Rangers" who could be appointed by county governments at a
monthly wage not to exceed $125.[15]

The county supervisors also were in annual session at Phoenix, and a
resolution was introduced to do away with the Rangers. The resolution
caused "a lively discussion," but when the supervisors deadlocked ten to
ten over the matter, it was decided to table the resolution and offer no rec-
ommendation about the Rangers to the legislature.[16]

Across the territory there were strong feelings about the Rangers.
Since the fall elections of 1908, Democratic newspapers had called for
abolition of the force.[17] There was considerable opposition to the company in the northern counties, since just one-third of the men were stationed in the "Northern Detachment" under the direction of Lieutenant Old. In earlier years the Rangers had almost completely neglected
the northern counties, mainly because there was far more crime in the
south. Now that the Rangers' presence was being felt in such communities as Winslow and Williams, their no-nonsense crackdowns may
have been resented. Although some resentment existed in the south
among those who had felt Ranger heavy-handedness, for the most part
southern Arizona enthusiastically supported the company of peace officers which had imposed order upon their unruly area. It long had
been rumored that many sheriffs were jealous of the Rangers, particularly since Ranger arrests often cost local officers fees and rewards. But
many sheriffs had cooperated successfully with the Rangers, and
Wheeler had collected letters of support from eight Arizona sheriffs and

*Seated in front of the Arizona capitol at far right, with a big hat on his knee, is Thomas Weedin. A Democratic councilman from Pinal County, Weedin introduced the 1909 bill to abolish the Rangers. He was strongly supported by fellow councilman Eugene Brady O'Neill, seated at far left. Seated third from left is newspaperman Kean St. Charles, a member of the Assembly in 1901 who was the only legislator to express opposition to the creation of the Rangers.*

several district attorneys. Mulford Winsor, a keen observer of Arizona events of the day and a knowledgeable Ranger historian, felt that "quite likely a majority" of Arizona citizens supported the Rangers.[18]

On Thursday, January 28, Captain Wheeler's report for December reached the governor's office. Newspapers supportive of the Rangers trumpeted that the company had made fifty-eight arrests during the month, fifty-five of which had resulted in convictions. It was emphasized that two Rangers had been sent to the mining camp of Courtland and promptly had enforced order upon the lawless population. The Tucson *Citizen* concluded that the work of the Rangers "has not conflicted with that of other peace officers but has been supplemental to it. The arrests made by them have been those which would not have been made but for them."[19]

But on Friday afternoon the Council voted to pass the bill to repeal the Rangers. The next day there was a joint caucus of the House and Council. The caucus voted overwhelmingly to abolish the office of public examiner, and even though Governor Kibbey had supported continuation of the office there was no opposition. On the Ranger issue, however, legislators from Cochise, Graham, and Pinal counties were outspoken in their arguments that the company was "a necessity in the Territory." [20] But the large Democratic majority held firm, and the Council voted to abolish the Rangers. It was regarded as certain that the Council vote insured that the House would support the anti-Ranger bill when it went before the floor the next week.

During these days Captain Wheeler had called the Rangers to Globe, but the members of the company were painfully aware of the legislative threat to their force. Ranger morale, of course, began to suffer, and there was a noticeable increase in criminal activities as lawbreakers opportunistically sensed the demise of their formidable adversaries. Captain Wheeler returned to Ranger headquarters from Globe, but almost immediately he had to ride into the field to search for horse rustlers. He instructed his headquarters clerk, Pvt. Emil Lenz, to write the governor's office that the respectable citizens of Naco supported the Rangers, and that Neill Bailey, one of three Cochise County representatives, had worked with Thomas Weedin to undermine the Rangers. Bailey, a Democrat from Naco, supposedly had supported the Rangers until the most recent elections, when Wheeler had interfered with some highly questionable practices. His ire thus aroused against the company, Bailey now had become a leader of the anti-Ranger movement. [21]

On the day the letter was written, February 3, the House of Representatives voted its approval of the anti-Ranger bill. Two days later Captain Wheeler rode in from his wilderness pursuit to Douglas, where he learned of the House vote. He wrote a long letter to the governor, expressing "the general and violent opposition" to the pending abolishment of the company. Wheeler was understandably dismayed that he had not been called to testify before the legislature about the Rangers. Complaining about the "false and malicious tactics [that] have been resorted to by Neil Bailey, among the Democratic members of the House and Council," Wheeler stated that he had telegraphed Ben Goodrich, a councilmember from Tombstone, to have him "summoned before any Committee." Wheeler asked the governor's aid in having him brought to Phoenix so that he could "answer freely and willingly, each and every question" that might occur to the legislators about the Rangers. [22]

Wheeler then made a startling proposal. So that no one could attribute his defense of the Rangers to a selfish desire to perpetuate his captaincy, he intended to resign his commission. "The pride of my life has been my Company," he said. But he felt that his resignation, following an impassioned testimony before the legislature, might persuade enough "fair minded men to turn the tide in favor of the service." Wheeler was willing to make this personal sacrifice for the good of the Rangers and the

people of Arizona. The sacrificial offer was publicized in Arizona newspapers, and on Thursday, February 11, the governor's office asked the
captain to come to Phoenix. Even though he was on the scene, no invitation to testify was sent from either house.[23]

Letters of protest began to pour into the governor's office and into
Ranger headquarters. Ranchers, mining executives, law officers, government officials, merchants, and Rangers themselves decried abolishment of
the company. The Tucson Chamber of Commerce sent letters to the Pima
County delegation and to other legislators denouncing the abolishment of
the force. The Rangers were widely praised as the most important element
in subduing outlawry in Arizona, and the governor was urged to "do all in
your power to retain our ranger force." [24]

John H. Slaughter, owner of the San Bernardino Ranch and legendary sheriff of Cochise County during the 1880s, issued a newspaper statement praising the Rangers in enthusiastic detail. Another lengthy newspaper endorsement came from the first captain of the Rangers, Burt
Mossman, who pled for continuation of the force he had founded, even if
economy demanded reduction to the original thirteen men. Harry
Wheeler himself, despairing of being called to testify, contributed a long
interview to the newspaper skirmish. He detailed Ranger services to Arizona, commented upon the relatively inexpensive cost of the force, emphasized the support he had received from various Arizona sheriffs and other
citizens, and otherwise stated his case before the public.[25]

There also was considerable editorial support for continuation of the
Rangers. Responding to journalistic criticism in the *Citizen*, Democratic
Councilman James Finley of Tucson defended his anti-Ranger vote in a
letter to the editor, denying that the caucus had influenced him and offering standard arguments as his justification. Finley objected to Rangers
serving as livestock inspectors, insisting that such expenses should be
borne by cattlemen rather than taxpayers. Finley also pointed out that the
Rangers "have been devoting much of their time to looking after smuggling on the border, the expense of which should have been borne by the
Federal government." Rangers had indeed been active as livestock inspectors, and for years they had patrolled the border to assist line riders of
Mexico and the United States in controlling smugglers, although much of
this effort had curtailed rustling as well. The *Citizen* fired back that since
opinion in favor of retaining the Rangers "is almost unanimous" among
his constituents, that Finley was duty-bound to reverse his position and
"represent the large majority of the people he was elected to represent." [26]

It was anticipated that upon the return of Governor Kibbey from
Washington, he would veto the bills abolishing the Rangers and the office
of public examiner. Everyone expected the legislature to override the governor's veto regarding the public examiner, but the avalanche of public

protest raised hope in some quarters that the Democratic caucus might relent on behalf of the Rangers.[27]

Kibbey was back in Phoenix by the end of the week, and he prepared a lengthy message to the assembly. On Monday morning, February 15, the governor vetoed the bills which abolished the Rangers and the office of public examiner. The bills were returned to the legislature along with the veto message, arriving in the Council chamber a little after 10:00 A.M.

The Council had just convened and only a small crowd was looking down from the ornate, three-sided gallery. As soon as the reading of the minutes was completed, the message regarding the Ranger bill was deliberated. It took more than half an hour to read this message. In his carefully worded communication, Kibbey repeated the arguments he had stated in support of the Rangers at the opening session of the legislature. Then he stated that he had received letters endorsing the Rangers from numerous sheriffs and district attorneys, and he quoted long passages from two letters. Next he took up the various criticisms proffered by Ranger opponents and refuted them, point by point. His message concluded with a long attack upon the political bias that had made the Rangers a Democratic target.[28]

At one point during the message, as Kibbey criticized the "inconsistent attitude" of legislators who once had supported the Rangers, the reading was interrupted by Brady O'Neill, a strong critic of the governor. O'Neill moved that the further reading of the message be dispensed with, and a few fellow Democrats clapped their hands in support. But the presiding officer, George Hunt, ruled against the motion. Hunt stated that if the Council intended to vote upon the governor's veto, they should have the courtesy to hear his remarks.

The reading continued, and when the veto message finally was finished Hunt pronounced the customary question: "Shall the bill pass the governor's objections thereto, notwithstanding?" The Council members had expected the vetoes, and the caucus had determined to remain firm. All ten Democrats voted against the veto, leaving only two Republicans to support the governor.

Thomas Weedin, who had originated the anti-Ranger bill, asked to be heard on a question of personal privilege. Granted the floor, Weedin launched a thinly disguised attack upon the governor. He recalled that two years earlier he had introduced a similar bill, and he reiterated his anti-Ranger arguments of 1907. Referring to the governor's assertion that the Rangers were necessary to preserve law and order, Weedin claimed that this was a reflection upon other Arizona peace officers, indeed, upon Arizona itself. "I resent the slander," declared Weedin vehemently.[29]

Vigorous applause erupted. Weedin then charged that each Ranger had been forced to contribute ten percent of his salary to Republican campaign chests. This erroneous claim was followed by an illusory excursion into mathematics: Weedin stated that the Rangers had cost Arizona $66,000 per year (during the previous fiscal year Ranger expenses totaled $28,476.31),[30] while there had been 1,100 arrests — and each arrest supposedly had cost the territory more than $3,000.

William J. Morgan, Democrat from Navajo County, claimed that every man, woman, and child in his county supported his anti-Ranger vote. Morgan went on to praise the stellar qualities of the local peace officers of Navajo County. Brady O'Neill then blurted out a coarse insult against the governor. Several people in the gallery hissed their disapproval, and Hunt pounded his desk. But O'Neill continued to snarl vituperation against Kibbey. His remarks became personal as he criticized Kibbey's record "and flayed him unmercifully." [31]

Captain Wheeler, still hoping in vain to be asked to testify on behalf of his beloved company, looked on with suppressed anger as the Rangers were vilified. The steel-nerved man of action watched helplessly as the Rangers met doom at the hands of partisan politicians.

The Council finally returned to business and overrode the governor's veto of the bill abolishing the public examiner's office. Word quickly spread that in the Council the necessary two-thirds majority had been reached to abolish the Rangers and the public examiner, and when the House convened at 1:00 a large crowd had gathered in the gallery. The messages were immediately read, and one after another the two vetoes were voted down. One Democrat strayed from the fold on the Ranger bill, but the 17 to 7 vote was more than Ranger opponents needed to abolish the force. Speaker of the House Sam F. Webb added his signature to that of George W. P. Hunt. Section 3 of the anti-Ranger bill stated: "This act shall be in force and effect from and after its passage." There was to be no transition period.[32]

Early in the afternoon of February 15, 1909, the Arizona Rangers abruptly ceased to exist. That day Rangers apprehended horse thief Peter Morris in the Chiricahua Mountains, and on a ranch in the Dragoon Mountains Pvt. Emil Lenz traded shots with Allen Rose before the robber would submit to arrest.[33]

Harry Wheeler returned to Naco, still seething with resentment. Each of the Rangers had to be notified that the company no longer existed, but several of the men were in the field and could not be reached for a time. Pvt. John McK. Redmond, searching remote country for horse thieves, tried to close with outlaws for five days after the company had been abolished. The rustlers, trying to stay ahead of Redmond, were captured by another officer. "It would have been a pretty 'how-dy-do'," growled Redmond, "if I had been killed or had killed one of them after my commission had been taken from me." [34]

On February 17 Wheeler forwarded to the governor's office the discharge papers, in duplicate, of the members of the final roster. He

*Pvt. John McK. Redmond, who continued to chase horse thieves after the Rangers were abolished.*

awarded "Excellent Discharge" to seventeen of his men. Typically, Wheeler downgraded the discharge of Tom Gadberry, "too recent in the force for me to know him intimately," and that of H. E. Woods, who, when summoned to Globe, "allowed him self to be detained by such a thing as a swollen stream of water." But the next day Wheeler relented, stating in a follow-up letter that he felt that it would be an "injustice" to the long faithful Woods to grant him less than an excellent discharge.[35]

The following day Wheeler accumulated the materials his men had sent in, and he compiled the monthly report for January 1909. Before detailing the arrests, the captain described the heroic way in which his men had overcome obstacles of water and distance in hurrying to Globe to avert labor difficulties. He mentioned that Billy Speed once had patrolled 754 miles horseback during one month, the all-time Ranger total, and Wheeler otherwise praised the efforts of his "faithful and loyal body of Public Servants." The seventeen-page report is handwritten; the ribbon on the headquarters' Oliver typewriter had worn out, and Wheeler could find no replacement.[36] All of Wheeler's remaining correspondence was composed in his bold handwriting. Although Wheeler now was off the territorial payroll, his sense of duty and deep affection for his job kept him at headquarters to finish the tasks necessary to tie up the loose ends of his company.

During the latter half of February, a few letters arrived at the governor's office offering support if some method could be found to reinstate the company. Wheeler released a letter refuting the criticisms of Thomas Weedin and other Ranger opponents, and he expressed the gratitude of the Rangers to the citizens who had supported the company.[37]

On Thursday, March 5, Harry Wheeler wrote the report of Ranger operations during the first fifteen days of February. The men in the field had ridden an average of thirteen and one-half miles per day, and forty-one arrests had been made. Wheeler proudly pointed out that not one arrest "was acquitted, compromised nor dismissed." He reported that sixteen head of cattle and three horses and saddles were recovered from a hideout in the Huachuca Mountains. Rangers had been working on this case for months, and arrests would have followed if the company had not been repealed.[38]

Now Harry Wheeler could linger at his beloved occupation no longer. Although his discharge would not be sent by the governor — accompanied by a highly complimentary letter of gratitude — until March 25,[39] Captain Wheeler had no further functions to perform. The last Ranger left his post.

# Epilogue

*"In those days all we had was dangerous work."*
— Pvt. John R. Clarke

The Rangers had performed their work well. Contrary to the dire predictions of various Ranger supporters, general outlawry did not return to Arizona. Civilization spread rapidly throughout the territory, and in 1912 Arizona achieved statehood.

By that time Harry Wheeler was sheriff of Cochise County. He had served briefly as a Cochise County deputy sheriff in 1909, then became a line rider for the customs service out of Douglas. He was elected sheriff by Cochise County voters in a special election in December 1911, and he won reelection in 1914 and 1916. Wheeler proved to be an industrious and innovative sheriff. His most notable action as sheriff came during a labor dispute in Bisbee in June 1917.[1]

As a Ranger, Wheeler had been assigned more than once to restore order during labor agitation, and he readily sided with the management when I.W.W. leaders triggered a strike at Bisbee's huge Copper Queen Mine. Sheriff Wheeler had been trying to join the army ever since the United States entered into the Great War in April 1917, and he regarded it as unpatriotic to deprive the war effort of Bisbee copper. Arbitrarily, he appointed 275 special deputies and began rounding up I.W.W. "wobblies" and suspected sympathizers of the strikers. On July 12, 1917, 1,187 agitators and miners, mostly aliens, were loaded onto cattle cars and taken by train to the hamlet of Hermanas, a short distance from the New Mexico line. The deportees were dumped off the train, although food and tents were sent from a nearby military post.[2]

Wheeler assumed complete responsibility for the "Bisbee Deportation," which caused a considerable uproar. He was charged with kidnaping, but the district attorney would not prosecute the popular sheriff. In March 1918 Sheriff Wheeler resigned to enlist in the army. Finally commissioned a captain after frantic attempts to get into the war, the forty-one-year-old cavalry veteran was on a troop ship for France by April. But before he could see combat duty, Captain Wheeler was called back to Arizona to face further court action resulting from the Bisbee Deportation. Detached to Fort Huachuca while the trial dragged on (eventually all defendants were acquitted), Wheeler was given an honorable discharge in December 1918.[3]

Soon Wheeler obtained a position as special officer for the Douglas police. Wheeler's son, Allyn, had died in 1915 at the age of seventeen following an automobile accident, and in 1919 Harry divorced his wife of twenty-one years. Less than two months later, Wheeler married an eighteen-year-old resident of Douglas who bore him three children over the next few years. He tried his hand at cattle ranching, but drought drove him out of business inside a year. In 1922 he ran for a fourth term as sheriff of Cochise County, but the deportation notoriety had damaged his popularity and he was defeated by more than 600 votes in the Democratic primary. He settled in the Bisbee area and tried to develop mine leases. Still a superb marksman, he was a charter member of the Tombstone Rifle Club, won high individual gun in Arizona in 1919, and repeatedly scored well in state and national competitions.[4]

He fell ill on Friday, December 11, 1925, and was taken to the Calumet and Arizona Hospital. Pneumonia set in, and Wheeler died on Thursday, December 17. The Palace Funeral Parlor (which had handled the body of Jeff Kidder in 1908) took charge of the remains, and Capt. Harry Wheeler was buried with military honors. He was forty-nine.[5]

After leaving the Rangers in 1902, Burt Mossman was unemployed for several months, although he spent considerable time being shown the bright lights of New York City by Col. W. C. Greene. Finally, he returned to ranching, operating the vast Diamond A lease in South Dakota. In 1905, at thirty-eight, Mossman married in Kansas City during a brief stopover on a journey to New York. The next year a son was born and in 1909 Grace Mossman gave birth to a daughter; however, the young mother died nine days later. In 1916 Mossman, still presiding over the expanding Diamond A. Cattle Company, moved permanently to Roswell, New Mexico.

Burt remarried in 1925; his bride, Ruth Shrader, was thirty-three, while Mossman was a vigorous fifty-eight. During the 1930s, however, Mossman began to suffer severely from arthritis. In 1943 his son, Maj. Burton C. "Billy" Mossman, Jr., died in a plane crash. Burt retired the next year at seventy-seven. He lived until 1956, although arthritis confined him to a wheelchair during his latter years.[6]

Tom Rynning stayed on as warden of the new prison at Florence until 1912, when a new Democratic administration replaced Arizona Republicans in various positions. Rynning moved to San Diego, but when Republicans returned to power in 1921 he was reappointed warden and moved back to Florence with his wife Margaret and three daughters. In 1932 he returned to California for good; two years later he was commissioned a deputy U.S. marshal for the San Diego division, and he also served as an undersheriff. At the age of seventy-five, in 1941, Rynning was peacefully tending flowers in the garden at his San Diego home when he was struck by a fatal heart attack.[7]

Rye Miles died the next year. After his Ranger days, he moved his wife and six children to Casa Grande, where he served for several years as a Pinal County deputy sheriff. He was the constable of Benson for a time, and ironically he spent a couple of years as a Cochise County deputy Ranger. Later he put in a long stint as a guard at the Arizona State Prison. During the thirty years that he wore one badge or another, he killed four men. Miles passed away quietly at the age of seventy-six.[8]

Many other ex-Rangers remained in law enforcement. Joe Pearce spent six years as a line rider; he also was chief of Apache police on the Fort Apache Reservation. An expert trailer, he became known as "Lone Wolf" to the Indians because he preferred to operate alone. He married and raised nine children. After he stopped wearing a badge, he ran a spread near Mount Baldy in Apache County. He had a nine-room house, an orchard, acreage under cultivation, and a herd of about 100 head of cattle. He served as vice-president of the Arizona Pioneers Historical Society, and in 1949, when he was seventy-five, Pearce applied for the position of police chief of Evants, Kentucky, in the heart of feuding territory, bloody Harlan County. When he sent in his application he declined, contrary to specifications, to send recommendations: "Am not mailing you any outside endorsements as the mere fact of my being a candidate should be sufficient." When the application of the elderly ex-Ranger attracted comment, Pearce shrugged. "Shucks," he said, "varmints is varmints and I'd just as lief plug 'em back in Kentucky as anywhere else, providin' the grub is there." But the candidacy of the old western gunfighter was declined, probably to the loss of Evants. The whimsical Pearce tried again, however, applying in 1954 for membership in the U.S. Border Patrol. On his application he stated his age as "past 21" (he was eighty). He died serenely in 1958.[9]

After Bud Bassett left the Rangers in 1905, he was a prison guard at Yuma, cowboyed and farmed for a while, then pinned on deputy's badges in Maricopa, Cochise, Mohave, and Yavapai counties. For years he was a state cattle inspector, and his final job was as night watchman for the Arizona Highway Department (at seventy-four he was the oldest active law officer in the United States). Bassett's career was capped by being appointed an honorary deputy sheriff of all fourteen Arizona counties.[10]

Oscar Felton was hired by the Globe police force. During a midnight patrol in November 1907, Felton was gunned down in the street and a bullet lodged in his leg. But the ex-Ranger emptied his revolver at his attacker, drilling Jack Nelson twice in the chest, once in the abdomen, and once in the shoulder.[11] Luke Short became a customs inspector, and in 1914 he shot and killed a malefactor. He was the first peace officer elected in the mining camp of Paradise. In 1912 Short moved to Douglas, serving still as a customs officer, then with the U.S. Department of Justice during World War I. He reared four children, then died in 1929 at the age of fifty-four.[12] Reuben Neill, after four years as a Ranger, served as a Cococino County deputy for fourteen years, and he also was chief of police at Flagstaff and Winslow. In all, Neill wore a badge in Arizona for thirty-five years.[13]

After Lt. W. D. Allison left the Rangers in 1904, he hired out as W. C. Greene's security man in Cananea. Following the riots of 1906, Allison moved to Texas and again pinned on a badge. His most famous encounter occurred when he intercepted cattle rustler Pascual Morosco, formerly a general under *Presidente* Francisco Madero. There was a hard fight, but Allison killed Morosco. In 1923, however, at the age of sixty-two, Allison was shot to death in Seminole, Texas.[14]

Dayton Graham, who briefly had served as the first sergeant of the Rangers, also had the old-time gunfighter's inclination to engage in shooting scrapes. Wounded in a 1903 Douglas saloon fight, which resulted in the death of another officer, Graham sought out his assailant a few days later and shot him to death. In another fray he severely wounded a Mexican who tried to resist arrest. Graham was killed at a sawmill at Casa Grande.[15]

It was rumored among Southwesterners that after the Rangers were disbanded, Lt. Billy Old slipped into Mexico to hunt down and kill the men responsible for the death of his friend, Jeff Kidder. Later, Old was mortally wounded by a woman in New Mexico.[16]

An old Ranger nemesis, Burt Alvord, managed to elude final capture by the Rangers or any other officers. He disappeared from Arizona, after which various reports had him turning up in Jamaica, Canada, Venezuela, Panama, and Honduras. Two stories suggested that Alvord died in 1910, either in Honduras or on a small island off the coast of Panama.[17] Alvord's *compañero*, Billy Stiles, left Arizona for Nevada, became a deputy sheriff, and in 1908 "perished as he lived, violently." Deputy Sheriff "William Larkin" (Billy's full name was William Larkin Stiles) was instrumental in the death of a Nevada cattle thief, but as a consequence the rustler's brother caught Billy with his back turned and triggered a fatal bullet into the former Arizona Ranger and outlaw.[18]

In 1955 Arizona belatedly demonstrated official solicitude for the Rangers by authorizing a $100 monthly pension to anyone who had served for six months and still lived in Arizona. Only five men qualified, although a few other surviving Rangers lived outside of Arizona.[19]

During the heyday of television Westerns, when every imaginable concept was produced as a series, ABC turned to the adventures of the Arizona Rangers. *26 Men* debuted in 1958, starring Tris Coffin, a tall, mustached, silver-haired actor of distinguished mien. Coffin, who had appeared in a dozen Western movies — often as a villain — portrayed Capt. Tom Rynning. But the televised exploits of the Rangers did not catch the public's imagination, and the series was canceled after one season.[20] Despite the exciting history of the Rangers and the long popularity of Western themes in movie houses, only one motion picture was ever lensed about the Rangers. In 1948 RKO released *Arizona Ranger*, a black and white, low-budget film that ran as a programmer on double-feature bills

*When* 26 Men *debuted on television in 1958, actors mingled with elderly ex-Rangers. Tris Coffin, the tall, silver-haired thespian who played Capt. Tom Rynning, and Kelo Henderson, who portrayed fictional Ranger "Clint Travis," stand behind real-life Rangers. Left to right in front are John McK. Redmond, Chapo Beaty, Joe Pearce, and Oliver Parmer.*

in small-town theaters. The movie starred Tim Holt and his father Jack Holt, a longtime leading man in action films. Tim, unsettled after his discharge from the Rough Riders, clashes with his screen father — played by his real father — and joins the Arizona Rangers. The pattern of an ex-Rough Rider enlisting in the Rangers was, of course, characteristic, but *Arizona Ranger* was a minor film which generated no further cinematic interest in the adventures of the Rangers.[21]

Occasionally, some of the Rangers got together and rehashed the wild old days. In 1940, for example, seven former members of the company — Oliver Parmer, O. C. Wilson, Jim Bailey, Charles A. Eperson, Rye Miles, Joe Pearce, and Charles McGarr — rode in the Prescott Rodeo Parade. They all brought hardware — six-guns and holsters and four old 1895 Winchesters — and they posed self-consciously for a group photo. And in 1958, when *26 Men* premiered on television, the actors traveled to Arizona and posed with four elderly Ranger survivors: John McK. Redmond, Chapo Beaty, Joe Pearce, and Oliver Parmer.

Chapo Beaty had ranched for several years in the Mustang Mountains after leaving the Rangers, but he lost his land in 1921. When he was fifty-eight he married eighteen-year-old Rita Rodriguez. "He called me 'honey'," she recalled fondly, adding, "I was good in history, too." The Beatys reared four sons in Patagonia. Chapo rode horses until three months before his death in 1964, at the age of ninety.[22]

Joe Pearce passed away in 1958, and so did John Foster, a former Ranger lieutenant who had moved to California in 1934. Oliver Parmer

*Seven ex-Rangers who rode in a 1940 rodeo parade. Left to right: Oliver Parmer, O. C. Wilson, Jim Bailey, Charles Eperson, Rye Miles, Joe Pearce, and Charles McGarr.*

died in 1961, while Timberline Bill Sparks, at age ninety-five thinking himself the last Ranger, expired in 1963 in Phoenix. John McK. Redmond died in 1967.[23]

The last surviving Ranger was John R. Clarke. As a twenty-two-year-old blacksmith in Nogales, he had handled himself impressively during a brawl and was observed by Billy Old. Old persuaded Clarke to enlist in the Rangers and taught him how to shoot. "In those days," reflected Clarke, "all we had was dangerous work." Clarke served two enlistments, 1906–1908, then wandered about doing oddjobs. He settled in California and died at his Rosemead home on May 5, 1982, at the age of ninety-seven. He was buried in Whittier.[24]

Arizona never forgot the Rangers, as the Olin Mathieson Corporation discovered in 1963. As part of their nationwide advertising campaign for Winchester rifles, a magazine ad showed the 1903 photograph of the entire Ranger company in a long line at Morenci. The photo was cropped, however, showing only nineteen men. But worse by far, the caption read: "19 Texas Rangers; only 18 Winchesters. Why?"[25]

A furor arose across Arizona. *Texas* Rangers indeed! Henry H. Hunter, director of public relations and advertising of the corporation, wrote to Arizona Governor Paul Fannin: "It became plain that we had rustled some of Arizona's history past and, unkindest cut of all, had given it to Texas." The corporation arranged a public apology ceremony for June 13, 1963. The ceremony took place in Phoenix and was attended by

*In 1968 a modern band of Arizona Rangers had their photo taken at Morenci, site of the famous 1903 Ranger group photo.*

elderly Chapo Beaty and John McK. Redmond, as well as the widow of Joe Pearce.

The ad man responsible, R. Scott "Black Ralph" Healy, was handcuffed and led by modern Arizona Rangers to a horse under a hanging tree. Healy, muttering, "I deserve it," was given a last drag from a cigarette, then he was placed on the horse and blindfolded. But Governor Fannin came through with a "dramatic last moment reprieve," permitting Healy to clean state guns as an "alternative sentence." [26]

Arizona Ranger volunteer organizations still carry on the tradition of Ranger service across the state. Companies exist in nearly every large community in Arizona. Membership requirements include a thorough background investigation. Applicants must be male residents of Arizona who are at least twenty-one years of age and of sound physical condition. New members are trained in the handling and shooting of firearms, and they must demonstrate the firearms proficiency requirements of their individual companies. Recruits are trained in first aid, rescue methods, search and seizure techniques, civil defense procedures, and hunter safety.

Unlike the Rangers of 1901–1909, the modern companies sport uniforms of western cut, with black hats, boots and string ties. They must supply their own firearms, and because many of these men are conscious traditionalists, a number of modern Rangers carry Colt single-action .45's. Like the old-time Rangers, they wear silver badges with five-point ball stars and the inscription "Arizona Rangers." And like the Rangers of Mossman and Rynning and Wheeler, the modern Arizona Rangers support the work of local and county peace officers.

# Appendix A

## *The Ranger Roster*

There were 107 men recorded on the Ranger rolls. Forty-four men (forty-one percent) were from Texas, while nine were natives of Arizona. John Clarke, Frank Ford, Don Johnson, and Eugene Shute, at twenty-two, were the youngest recruits, and fifty-five-year-old John Rhodes was the elder statesman among rookies. The median age among recruits was a mature thirty-three. Tom Gadberry, who enlisted fifteen days before the Rangers were disbanded, served the briefest tenure. The lengthiest enlistment was served by James T. Holmes, who signed up on September 2, 1902; Frank Wheeler joined eight days later, and both men were still Rangers when the company was abolished.

| Name (Highest Rank) | Age at Enlistment | Place of Birth | Tenure |
|---|---|---|---|
| Allison, William D. (Lt.) | 42 | Ohio | 1903–1904 |
| Anderson, Robert M. (Pvt.) | 36 | Tennessee | 1902–1908 |
| Baggerly, Roy (Pvt.) | 30 | Texas | 1906 |
| Bailey, James D. (Pvt.) | 33 | Kentucky | 1903–1905 |
| Barefoot, Fred S. (Pvt.) | 40 | Mississippi | 1901–1903 |
| Bassett, James H. (Pvt.) | 34 | Texas | 1902–1905 |
| Bates, W. F. (Pvt.) | 24 | Texas | 1907 |
| Beaty, Clarence L. (Sgt.) | 28 | Oklahoma | 1903–1907 |
| Black, Samuel C. (Pvt.) | 45 | West Virginia | 1908 |
| Brooks, John J. (Lt.) | 36 | Texas | 1903–1905 |
| Brooks, Ross (Pvt.) | 42 | Texas | 1904–1905 |
| Burnett, Reuben E. (Pvt.) | 41 | Texas | 1905–1906 |
| Byrne, Cy (Pvt.) | 41 | Ohio | 1907–1909 |
| Campbell, John E. (Sgt.) | 35 | Pennsylvania | 1901–1903 |
| Carpenter, William L. (Pvt.) | 41 | California | 1908–1909 |
| Chase, Arthur F. (Sgt.) | 31 | Illinois | 1908 |
| Clarke, John R. (Pvt.) | 22 | Arizona | 1906–1908 |
| Coffee, Garland (Pvt.) | 29 | Arkansas | 1905 |
| Davis, Wayne (Pvt.) | 28 | Arizona | 1906–1908 |
| Devilbiss, George W. (Pvt.) | 31 | California | 1904 |
| Doak, Boyd M. (Pvt.) | 39 | Texas | 1905 |
| Ehle, A. E. (Pvt.) | 29 | Arizona | 1907–1908 |
| Ensor, William (Pvt.) | 38 | Texas | 1903–1905 |
| Eperson, Charles A. (Pvt.) | 26 | California | 1903–1906 |

| | | | |
|---|---|---|---|
| Farnsworth, Clark H. (Pvt.) | 31 | Illinois | 1905 |
| Felton, Oscar (Pvt.) | 27 | Oregon | 1902–1905 |
| Ferguson, William F. (Pvt.) | 44 | Texas | 1903–1904 |
| Ford, Frank A. (Pvt.) | 22 | Kansas | 1907 |
| Foster, John (Lt.) | 34 | South Carolina | 1902–1907 |
| Foster, William K. (Pvt.) | 36 | New York | 1903 |
| Fraser, J. A. (Pvt.) | 28 | Texas | 1907–1908 |
| Gadberry, M. Tom (Pvt.) | 40 | Texas | 1909 |
| Graham, Dayton (Sgt.) | 43 | Ohio | 1902 |
| Gray, Henry S. (Sgt.) | 47 | California | 1901–1905 |
| Greenwood, John F. (Pvt.) | 45 | Texas | 1905–1906 |
| Grover, Herbert E. (Pvt.) | 33 | Kansas | 1901–1902 |
| Gunner, Rudolph (Sgt.) | 35 | Texas | 1907–1909 |
| Hamblin, Duane (Pvt.) | 38 | Utah | 1901–1902 |
| Hayhurst, Samuel J. (Pvt.) | 32 | Texas | 1903–1909 |
| Heflin, C. W. (Pvt.) | 30 | Texas | 1908 |
| Henshaw, Samuel (Pvt.) | 37 | Texas | 1903 |
| Hickey, Marion M. (Pvt.) | | | 1905–1906 |
| Hicks, O. F. (Pvt.) | 36 | Missouri | 1909 |
| Hilburn, J. R. (Pvt.) | 41 | Texas | 1904 |
| Holland, Thomas J. (Pvt.) | 28 | Texas | 1901–1902 |
| Holmes, James T. (Sgt.) | 30 | Denmark | 1902–1909 |
| Hopkins, Arthur A. (Sgt.) | 26 | Colorado | 1903–1906 |
| Horne, R. D. (Pvt.) | 26 | Pennsylvania | 1907–1908 |
| Humm, George (Pvt.) | 34 | Ohio | 1908 |
| Johnson, Don (Pvt.) | 22 | Texas | 1901–1902 |
| Jorgenson, Louis (Pvt.) | 32 | Utah | 1903 |
| Kidder, Jefferson P. (Sgt.) | 28 | South Dakota | 1903–1908 |
| King, Orrie (Pvt.) | 37 | Texas | 1908–1909 |
| Larn, William A. (Pvt.) | 33 | Texas | 1907 |
| Lenz, Emil R. (Pvt.) | 27 | New York | 1908–1909 |
| MacDonald, Alex R. (Sgt.) | 39 | Mauritius | 1903–1904 |
| McAda, Oscar (Sgt.) | 26 | Texas | 1907–1909 |
| McDonald, James Porter (Pvt.) | 37 | Texas | 1905–1907 |
| McGarr, Charles E. (Pvt.) | 35 | Arizona | 1905–1907 |
| McGee, E. S. (Pvt.) | 32 | Texas | 1907–1909 |
| McGee, James E. (Sgt.) | 34 | Arkansas | 1904–1906 |
| McKinney, Joseph T. (Pvt.) | 46 | Arkansas | 1905–1906 |
| McPhaul, Henry H. (Pvt.) | 38 | Texas | 1905–1906 |
| Mayer, George L. (Pvt.) | 41 | New York | 1907 |
| Mickey, Lew H. (Sgt.) | 37 | Nebraska | 1905–1909 |
| Miles, J. T. (Sgt.) | 41 | Texas | 1907–1908 |
| Moran, James (Pvt.) | 29 | California | 1904–1905 |
| Mossman, Burton C. (Capt.) | 34 | Illinois | 1901–1902 |
| Mullen, John Oscar (Pvt.) | 24 | California | 1903 |
| Neill, Reuben L. (Pvt.) | 28 | Arizona | 1904–1906 |
| Old, William A. (Lt.) | 32 | Texas | 1904–1909 |
| Olney, Benjamin W. (Pvt.) | 34 | Texas | 1906–1909 |
| Page, Leonard S. (Pvt.) | 25 | Texas | 1901–1902 |
| Parmer, William C. (Pvt.) | 25 | Texas | 1908–1909 |

| | | | |
|---|---|---|---|
| Pearce, Joseph H. (Pvt.) | 30 | Iowa | 1903–1905 |
| Pearson, Pollard (Pvt.) | 34 | Texas | 1902–1903 |
| Peterson, William S. (Pvt.) | 38 | Texas | 1902–1905 |
| Poole, Travis B. (Pvt.) | 28 | Texas | 1907 |
| Redmond, John McK. (Pvt.) | 24 | New York | 1908–1909 |
| Rhodes, John (Pvt.) | 55 | Texas | 1906–1908 |
| Richardson, Frank (Pvt.) | 40 | Oregon | 1901–1902 |
| Rie, Charles (Pvt.) | 31 | Pennsylvania | 1903–1904 |
| Robinson, McDonald (Pvt.) | 33 | Texas | 1901–1902 |
| Rollins, Jesse W. (Pvt.) | 34 | Utah | 1906–1907 |
| Rountree, Oscar J. (Pvt.) | 27 | Texas | 1903–1906 |
| Rynning, Thomas H. (Capt.) | 36 | Wisconsin | 1902–1907 |
| Scarborough, George E. (Pvt.) | 23 | Texas | 1901–1902 |
| Short, Luke (Pvt.) | 33 | Texas | 1908 |
| Shute, Eugene H. (Pvt.) | 22 | Arizona | 1905–1906 |
| Smith, James (Pvt.) | 32 | Texas | 1907 |
| Sparks, William (Sgt.) | 42 | Iowa | 1903–1909 |
| Speed, William S. (Sgt.) | 35 | Texas | 1906–1909 |
| Splawn, C. T. (Pvt.) | 31 | Texas | 1905 |
| Stanford, Tip (Sgt.) | 29 | Texas | 1903–1909 |
| Stanton, Richard H. (Pvt.) | 30 | New York | 1901 |
| Stiles, William L. (Pvt.) | 32 | Arizona | 1902 |
| Tafolla, Carlos (Pvt.) | 36 | New Mexico | 1901 |
| Thompson, Ray (Pvt.) | 30 | Nevada | 1905–1907 |
| Warford, David E. (Pvt.) | 36 | New York | 1903 |
| Warren, James (Pvt.) | 37 | Illinois | 1901–1902 |
| Webb, William W. (Pvt.) | 33 | Texas | 1902–1903 |
| Wheeler, Frank S. (Sgt.) | 31 | Mississippi | 1902–1909 |
| Wheeler, Harry C. (Capt.) | 26 | Florida | 1903–1909 |
| Wilson, Owen C. (Pvt.) | 29 | Texas | 1903 |
| Wilson, W. N. (Pvt.) | 35 | Arkansas | 1906–1909 |
| Woods, Herbert E. (Pvt.) | 27 | Arizona | 1908–1909 |
| Woods, Leslie K. (Pvt.) | 29 | Arizona | 1906 |

# Appendix B

## *Arizona Ranger Legislation*

REVISED STATUTES OF ARIZONA OF 1901

### CHAPTER II

### ARIZONA RANGERS

3213. (Sec. 1.) That the governor of this territory is hereby authorized to raise and muster into service of this territory, for the protection of the frontier of this territory, and for the preservation of the peace and the capture of persons charged with crime, one company of Arizona rangers, to be raised as hereinafter prescribed, and to consist of one captain, one sergeant and not more than twelve privates, each entitled to pay as follows: Captain to receive one hundred and twenty ($120.00) dollars per month; sergeant to receive seventy-five ($75.00) dollars per month; and privates fifty-five ($55.00) dollars per month, each; and the pay herein provided shall be full compensation in lieu of all other pay and compensation for clothing for both officers and men.

3214. (Sec. 2.) That the requisite number of officers and men for said company shall be raised, if possible, in the frontier counties of this territory.

3215. (Sec. 3.) That the governor is authorized and empowered, when in his opinion the public emergency shall require it, after the passage of this act, to appoint competent persons as captain and sergeant, and to enroll, as set forth in this act, the requisite number of men for the company; the captain shall return to the governor the muster roll and the report of the condition of the company, and the governor shall thereupon commission the said officers of said company, supply said company as under the provisions of this act he may deem proper and necessary, and order them upon duty in accordance with the provisions of this act.

3216. (Sec. 4.) Said men shall be furnished by the territory with the most effective and approved breech-loading cavalry arms, and for this purpose the governor is hereby authorized to contract in behalf of the territory for ten stands of arms, together with a full supply of ammunition, the same to be all of the same make and calibre, and each member of the company to be furnished with the arms to be used by him at the price the same shall cost the territory, which sum shall be retained out of the first money due him.

3217. (Sec. 5.) That each member of said company shall be required to furnish himself with a suitable horse, six shooting pistol (army size) and all necessary accoutrements and camp equipage, the same to be passed upon and approved by the enrolling officer before enlisted, and should any member fail to keep himself furnished as above required, then the officer in command shall be authorized and required to purchase the articles of which he may be deficient, and charge the cost of the same to the person for whom the same shall be provided: Provided, That all horses killed in action shall be replaced by the territory, and the cost of horses so killed in action shall be determined by the captain.

3218. (Sec. 6.) That said officers and men shall be furnished by the territory with provisions, ammunition, and forage for horses when necessary and when on duty.

3219. (Sec. 7.) Each member of said company is hereby authorized, when in pursuit of criminals, to take horses when necessary to continue the chase, wherever he may find them; said horses to be returned to the owners as soon as possible afterwards, and the same to be paid for by the territory.

3220. (Sec. 8.) That the men shall be enrolled for twelve months, unless sooner discharged, and at the expiration of their term of service others shall be enrolled to supply their places, in case the governor deems such action necessary for the protection of the frontier, or for the preservation of the peace, or the capture of persons charged with crime.

3221. (Sec. 9.) That no enlisted men shall be discharged from the service without special order from the governor, nor shall any member of said company dispose of or exchange their horses or arms without the consent of the commanding officer of the company while in service of the territory.

3222. (Sec. 10.) That the captain of the company shall use his own discretion as to the manner of operations, selecting as his base the most unprotected and exposed settlement of the frontier.

3223. (Sec. 11.) That the troops raised under and by virtue of this act shall be governed by the rules and regulations of the army of the United States, as far as the same may be applicable, but shall always be and remain subject to the authority of the Territory of Arizona for frontier service.

3224. (Sec. 12.) The captain of such company shall have authority to concentrate all of such company, or divide into squads for the purpose of following and capturing any outlaws, law breakers, marauding Indians, or bands of hostile Indians, or for the purpose of carrying out any measure that may contribute to the better security of the frontier; but the entire force raised under the provisions of this act, shall be at all times under and subject to the orders of the governor, and shall be exempt from all military, jury or other service; and that the intention of this act, with full power to remove any officer or man for incompetency, neglect of duty or disobedience of orders.

3225. (Sec. 13.) Members of said company shall have full power to make arrests of criminals in any part of the territory, and upon the arrest of any criminal, shall deliver the same over to the nearest peace officer in the county where the crime is committed.

3226. (Sec. 14.) It shall be the duty of the auditor of this territory to draw his warrant on the territorial treasurer at the end of each month for the pay of each officer and man in said company, and to forward the same to the captain of said company, and also a warrant for the amount of provision, ammunition and forage; but the food of each officer or man in said company shall not exceed in price the sum of one dollar per day and such forage shall not exceed the sum of fifty cents per day per horse; the same shall be forwarded upon the receipt by said officer of an itemized account from the captain of said company, to be signed by such captain and certified by him, and which shall be carefully scrutinized by such auditor, and should the same or any item therein be found unlawful or unreasonable, he shall suspend payment of the same and refer the same to the governor, who shall pass thereon and certify the same for the payment in such sum as he shall find correct and reasonable; and it shall be the duty of the territorial treasurer to pay such warrants out of the general fund as other warrants are paid.

3227. (Sec. 15.) That the captain shall be authorized to purchase all necessary pack animals to be furnished said company for transportation purposes, but not exceeding four in number; to purchase all necessary supplies to be delivered by contractors at the place to be designated by the captain of the company; and all accounts and certificates of such agent shall be examined and allowed by the captain of the company and certified by him, as the accounts for the payment of men, food or forage.

3228. (Sec. 16.) The governor shall have power to disband said company or any portion thereof when in his opinion their services shall no longer be necessary for frontier protection.

3229. (Sec. 17.) That there shall be annually levied and collected in addition to all other taxes authorized by law, a tax of five cents on the hundred dollars of taxable property in this territory to be placed in a fund by the territorial treasurer, to be known as the ranger fund, and upon which fund all warrants and payments made under any of the provisions of this act, shall be drawn and made. Said tax shall be levied and collected in the same manner, at the same time and by the same officers as other territorial taxes.

3230. (Sec. 18.) That no portion of said troops shall become a charge against this territory until organized and placed under orders.

(Took effect March 21, 1901.)

## ACTS, RESOLUTIONS AND MEMORIALS, 22ND LEGISLATIVE ASSEMBLY TERRITORY OF ARIZONA, 1903

No. 64.                                      AN ACT

To repeal Section 1, Paragraph 3213, and Section 5, Paragraph 3217, and Section 14, Paragraph 3226, and Section 15, Paragraph 3227, of Chapter II, Title 46, of the Revised Statutes of Arizona of 1901.

Be it enacted by the Legislative Assembly of the Territory of Arizona:

Section 1. That Section 1, Paragraph 3213, and Section 5, Paragraph 3217, and Section 14, Paragraph 3226, and Section 3227, of Chapter II, Title 46, of the Revised Statutes of Arizona, 1901, be and the same are hereby repealed and the following provisions be and the same are hereby enacted in lieu thereof.

Section 2. That the Governor of this Territory is hereby authorized to raise and muster into the service of this Territory and for the protection of the frontier of this Territory and for the preservation of peace and capture of persons charged with crime one company of Arizona Rangers, to be raised as hereinafter prescribed, and to consist of one captain, one lieutenant, four sergeants and not more than twenty (20) privates, each entitled to pay as follows:

Captain to receive one hundred and seventy-five ($175.00) dollars per month.

Lieutenant to receive one hundred and thirty ($130.00) dollars per month.

Sergeants to receive one hundred and ten ($110.00) dollars per month.

Privates to receive one hundred ($100.00) dollars per month.

And the pay herein provided shall be full compensation for all services rendered and expenses incurred of whatever character by the officers and men, except arms and ammunition, and the actual expenses of the captain in making out, rendering and transmitting the reports and statements required of him by law and for other necessary and incidental expenses, said expenses not to exceed twenty dollars ($20.00) per month, provided that, all members injured while in the performance of their duties shall have medical treatment and care at the expense of the Territory.

Section 3. That each member of said company shall be required to furnish himself with a suitable horse and pack animal, six-shooting pistol (army size), and all necessary accoutrements and camp equipage, the same to be passed upon and approved by the enrolling officer before enlisted and should any member fail to keep himself furnished as above required, then the officer in command shall be authorized and required to purchase the articles of which he may be deficient and charge the cost of same to the person for whom the same shall be provided; provided that all horses killed in action shall be replaced by the Territory, and the cost of horses so killed in action shall be determined by the captain.

Section 4. It shall be the duty of the Auditor of this Territory to draw his warrant on the Territorial Treasurer at the end of each month for the pay of each officer and man in said company, and to forward the same to the captain of said company, and also a warrant for the amount of ammunition and for the payment of expenses incurred in caring for any injured members and for necessary and incidental expenses, and the same shall be forwarded upon the receipt by said officer of an itemized account from the captain of said company to be signed by such captain and certified by him, and which shall be carefully scrutinized by such Auditor, and should the same or any item therein be found unlawful or unreasonable, he shall suspend payment of the same and refer the same to the Governor, who shall pass thereon and certify the same for the payment in such sum as he shall find correct and reasonable.

Section 5. That Paragraph 3227, Section 15, of the Revised Statutes of Arizona, 1901, be and the same is hereby repealed.

Section 6. That upon this Act taking effect the captain of the Rangers shall dispose of at public sale all pack animals and pack equipments now in possession of the Rangers, and shall forward the proceeds of said sale to the Territorial Treasurer for deposit in the Ranger fund, taking the Treasurer's receipt for same.

Section 7. That in case of a deficit in said Ranger fund the Territorial Auditor is hereby authorized and directed to draw his warrant on the general fund, and the Territorial Treasurer is hereby authorized and directed to pay the same as other warrants are paid.

Section 8. The captain shall provide and issue to each Ranger a badge, uniform in size and shape, with the words "Arizona Ranger" inscribed thereon in plain and legible letters, which badge shall be returned to the captain upon the said Ranger going out of service, the expense of which badge shall be paid for as part of the incidental expenses provided for.

Section 9. All Acts and parts of Acts in conflict with the provisions of this Act are hereby repealed.

Section 10. This Act shall take effect and be in force from and after its passage.

Approved March 19th, 1903.

## ACTS, RESOLUTIONS AND MEMORIALS, 25TH LEGISLATIVE ASSEMBLY TERRITORY OF ARIZONA, 1909

### AN ACT

Repealing the Act Establishing the Ranger Force of the Territory of Arizona.

Be it enacted by the Legislative Assembly of the Territory of Arizona:

Section 1. Act No. 64, Session Laws of the Twenty-Second Legislative Assembly of the Territory of Arizona, and all amendments thereto are hereby repealed.

Section 2. All Acts and parts of Acts in conflict with the provisions of this Act are hereby repealed.

Section 3. This Act shall be in force and effect from and after its passage.

This bill having been returned by the governor, with his objections thereto, having passed both houses by two-thirds vote of each house, has become a law this 15th day of February, A.D. 1909.

# Endnotes

## 1901: A New Force in Outlaw Territory

1. The Fairbank train robbery is described in the Tombstone *Epitaph* (February 15, 1900) and in Haley, *Jeff Milton*, 305–311.

2. The Gibbons-Lesueur murders are related in Jensen, "Birth of the Arizona Rangers," *Old West*, 30–33.

3. For details of the Stiles-Alvord escape read the Tombstone *Epitaph* (April 8 and 20, 1900). For Stiles's escape from Sheriff Lewis see the Nogales *Border Vidette* (February 2, 1901) in Peck memoirs.

4. Phoenix *Gazette*, quoted in *Border Vidette* (October 27, 1898).

5. Hunt, *Cap Mossman*, 143–145.

6. The best account of the creation of the Rangers was related by contemporary observer and Ranger historian Mulford Winsor, "The Arizona Rangers," *Our Sheriff and Police Journal*, 49–50.

7. The act creating the Rangers is in *Revised Statutes, Arizona Territory, 1901*, Chapter II, "Arizona Rangers," 833–836. See Appendix B.

8. Frazier Hunt consulted Mossman but added embellishments in writing *Cap Mossman, Last of the Great Cowmen*. Other biographical accounts of Mossman may be found in Coolidge, *Fighting Men of the West*, 247–279, and Raine, *Famous Sheriffs and Western Outlaws*, 236–252.

9. Mossman's meeting in the governor's office is related in Hunt, *Cap Mossman*, 145, and in an interview with Mossman, *Arizona Daily Star* (January 9, 1947). Mossman later recalled that his income prior to becoming Ranger captain was $1,000 per month (Hunt, *Cap Mossman*, 144). It seems unlikely that a man of Mossman's accustomed income and style of living would accept a position for a fraction of his usual revenue. Perhaps one or more of the wealthy men who persuaded Mossman to head up the new force assured him of a supplementary income, much like college football coaches whose comparatively modest institutional salaries are boosted by pay from oil companies or other private firms owned by supportive alumni.

10. Bisbee is described in: Bailey, *Bisbee: Queen of the Copper Camps;* Burgess, *Bisbee Not So Long Ago;* Chisholm, *Brewery Gulch;* and Wentworth, *Bisbee With the Big B*.

11. For enlistment dates and similar details see the Ranger personnel records at the Arizona State Archives, Phoenix.

12. The Ranger scuffle in Naco is described in the Bisbee *Review* (November 16, 1901).

13. Mossman related his interview with Governor Murphy about the excessive number of Republican recruits in the *Arizona Daily Star* (January 9, 1947).

189

14. Mossman's acquaintance with the 1895 Winchester is described in Hunt, *Cap Mossman*, 121–122. Also see Ernenwein, "Lucky Star," *Ranch Romances*, 44. An excellent photographic description of this rifle may be studied in Rattenbury, "A Portfolio of Firearms," *American West*, 44.

15. Ranger financial details may be found in the Arizona State Archives.

16. The account of the first big Ranger fight is based on: *Arizona Daily Citizen* (October 14, 1901); Pearce, "The Killing of Arizona Rangers at the 'Battle Ground'," *Arizona Stockman*, 7–9; Winsor, "The Arizona Rangers," *Our Sheriff and Police Journal*, 51–52; Ernenwein, "Lucky Star," *Ranch Romances*, 44–45; Sharp, "The Maxwells of Arizona," *Frontier Times*, 34–35; Hunt, *Cap Mossman*, 152–159.

17. Winsor, "The Arizona Rangers," *Our Sheriff and Police Journal*, 51.

18. Hunt, *Cap Mossman*, 153.

19. *Ibid.*

20. Ernenwein, "Lucky Star," 9.

21. Solomonville *Bulletin*, quoted in Miller, *Arizona Rangers*, 39.

22. Mrs. Tafolla's pension was finally approved by the legislature on March 19, 1903. See *Resolutions and Memorials, Twenty-second Legislative Assembly, Territory of Arizona, 1903*, No. 57:93.

23. Arrest totals are detailed in Report of Captain Burton Mossman of Arrests Made by Arizona Rangers from October 2, 1901 to July 30, 1902, available at the Arizona Heritage Center and in the Ranger archives.

24. The Head and Williams arrest is described in issues of the Tucson *Star* and Bisbee *Review*, quoted in Miller, *Arizona Rangers*, 40–41.

### 1902: Tracking Chacón and On to Douglas

1. On December 1, 1901, McDonald Robinson enlisted, replacing Tafolla and raising the Ranger complement to thirteen. But on December 2, 1901, Richard Stanton, out of favor since the November altercation with fellow Rangers, was discharged, reducing the roster to twelve and making room for Graham.

2. For an account of Graham's posse see Tucson *Citizen* (January 13, 1902). Graham's enlistment and reenlistment are detailed in his personnel file.

3. Ranger arrest activities during this period are detailed in Report of Captain Burton Mossman of Arrests Made by Arizona Rangers from October 2, 1901 to July 30, 1902. Mossman slipped into Mexico to arrest Walter Tremble, who six months earlier had raped a thirteen-year-old girl. The girl was the adopted daughter of a Graham County deputy sheriff named Hill who traced Tremble to Cananea. Informed of the situation, Mossman deftly handled the arrest and delivered the rapist and his brother to Hill and Sheriff Jim Parks in Bisbee. Tucson *Citizen* (February 27, 1902).

4. A description of the posse pursuit and arrests of Neill, Cook, and Roberts may be found in the Tucson *Citizen* (March 18, 1902) and in a letter from Pollard Pearson to Mulford Winsor (n.d.).

5. Interview of Mossman by Will C. Barnes (April 17, 1935).

6. Pollard Pearson to Mulford Winsor (n.d.).

7. Governor Murphy's praise of the Rangers is quoted from *Report of the Governor of Arizona, 1902*, "Arizona Rangers," 88.

8. For the fight resulting in the death of rustler Manuel Mendosa, see *Arizona Republican* (April 8, 1902).

9. Mossman's arrest report details the Ranger apprehensions of this period,

while the quotation praising the Rangers is in the *Border Vidette* (August 30, 1902), in Peck memoirs.

10. Bailey, *Bisbee: Queen of the Copper Camps*, 94.

11. The Ranger brawl with Bisbee policemen is described in the Tucson *Citizen* (August 19 and 21, 1902); in Egerton, "A Tale of Two Petitions," *Arizona Republican* (February 6, 1983); and in Bailey, *Bisbee: Queen of the Copper Camps*, 94–95.

12. *Arizona Republican* (November 22, 1922).

13. For Chacón's murder of Becker see Walters, *Tombstone's Yesterday*, 226, and Mossman's interview by Will C. Barnes (April 17, 1935).

14. See the *Border Vidette* (September 13 and 20, 1902) in Peck memoirs.

15. Mulford Winsor, who moved to Arizona in 1892 and who was deeply involved in the affairs of the territory, stated that when Mossman was commissioned captain, "a warrant for Chacón was placed in his hands." Winsor, "The Arizona Rangers," *Our Sheriff and Police Journal*, 52.

16. Alvord's early life in Tombstone is outlined in the Tucson *Citizen* (January 7, 1904), and his revolver practice with a string and can is described in Coolidge, *Fighting Men of the West*, 168.

17. The daring escape from Tombstone is described in the Tombstone *Epitaph* (April 8, 1900).

18. The location of Stiles's wife is mentioned in the Tucson *Citizen* (March 7, 1904).

19. Stiles to E. P. Drew, quoted in the *Border Vidette* (February 2, 1901), Peck memoirs.

20. Pegleg Smith quoted in the *Border Vidette*, 356, in Peck memoirs.

21. Mossman's interview with Will C. Barnes (April 17, 1935).

22. This account of Mossman's efforts to capture Chacón is based primarily on Mossman's words, as dictated to his close friend and fellow cattleman, Judge Will C. Barnes, on April 17, 1935. Another Barnes transcript is quoted in Winsor, "The Arizona Rangers," *Our Sheriff and Police Journal*, 53–54. Also see the Tucson *Citizen* (September 4, 1902).

23. Stiles's job as an ore train driver is reported in the Tucson *Citizen* (January 16, 1902).

24. An Arizona Ranger voucher, dated "January, 1902" and signed by "Wm Stiles," details his daily travels from January 15 to January 30:

| | | | |
|------|---------|-------------|--------|
| 15th | 3 meals | Naco | $1.00 |
| 16th | 3 meals | Naco | 1.00 |
| 17th | 3 meals | Naco | 1.00 |
| 18th | 3 meals | Naco | 1.00 |
| 19th | 1 meal | Naco | .35 |
| 19th | 2 meals | Tucson | .70 |
| 20th | 3 meals | Casa Grande | 1.00 |
| 20th | 3 meals | Casa Grande | 1.00 |
| 21st | 1 meal | Casa Grande | .35 |
| 21st | 2 meals | Vekol Mine | 1.00 |
| 22nd | 1 meal | Vekol Mine | .50 |
| 22nd | 1 meal | Vekol Mine | .50 |
| 23rd | 2 meals | Gun Site | 1.00 |
| 23rd to | | | |
| 30th | 2 meals | Mexico | 11.00 |
| | | | $21.40 |

Did Stiles err on the 20th, or did he perhaps feed an informant in Casa Grande?

25. Arcus Reddoch, quoted in the *Border Vidette*, 340, Peck memoirs.

26. *Ibid.*, 353.

27. Stiles's Ranger salary was turned in as Arizona Rangers Claim No. 146. This claim was signed by Captain Mossman and paid by Warrant No. 110.

28. Chacón's execution is described in: *Arizona Republican* (November 22, 1902); Bisbee *Review* (November 22, 1902); Tucson *Citizen* (November 21, 1902); and numerous other newspapers of the day.

29. Alvord's favorable treatment is described in the Tucson *Citizen* (December 2, 3, and 11, 1902). Legal pressure against Alvord and Stiles is described in the Tucson *Citizen* (March 3, July 16 and 29, December 7, 1903), and in Ball, *United States Marshals*, 224–225. The impoundment and sale of the outlaws' guns and saddles was related to me by Wally Zearing in an interview in Benson on March 13, 1983. Zearing bought five of the guns and later sold the weapons to a Sonoita collector.

30. Mossman's activities just after Chacón's arrest and his reasons for resigning are best explained in Will C. Barnes's 1935 interview.

31. Rynning's background is related in detail in the first 200 pages of his autobiography, *Gun Notches*. *Gun Notches*, however, must be used with extreme care. It is well to bear in mind Ranger Sam Hayhurst's evaluation: "Tom Rynning was a fine upstanding fellow and no bragger. I just can't understand how he ever allowed that book, 'Gun Notches', to go into print. It is full of mistakes. I know for I was with Rynning much of the time." "Reminiscences of Sam J. Hayhurst," as told to Mrs. George F. Kitt (September 27, 1937), 2.

32. Rynning's commission is detailed in his personnel record and in *Report of the Governor of Arizona, 1903*, "Arizona Rangers." In *Gun Notches* (189) Rynning commented that "I was serving as Captain of the Arizona Rangers at [President Roosevelt's] request," and he also stated that "President Roosevelt and Governor Brodie both insisted on me taking charge and reorganizing the Rangers" (203).

33. Rynning, *Gun Notches*, 205.

34. Pearce, "Line Rider," 103.

35. Ranger headquarters in Douglas is described in Pearce, "Line Rider," 105. The date of the move is given in *Report of the Governor of Arizona, 1903*, "Arizona Rangers."

36. Enlistment details and recruit backgrounds are given in the Ranger personnel records.

37. The background and quotation on James T. Holmes are from Arcus Reddoch, quoted in the *Border Vidette*, 349 of Peck memoirs.

38. Teddy Roosevelt liked to keep up with his former Rough Riders. Maj. W. H. H. Llewellyn, appointed U.S. attorney in New Mexico Territory, kept Roosevelt informed about the adventures of ex-Rough Riders in the Southwest. For example, the president received a somewhat inaccurate account of a shootout at Douglas involving Private Webb of the Arizona Rangers. Delighting in such tidbits, he wrote to Secretary of State John Hay: "Also have to report that Comrade Webb, late of Troop D [actually Troop B], has just killed two men at Bisbee, Arizona. Have not yet received the details of our comrade's trouble . . . but understand that . . . he was entirely justified in the transaction." Roosevelt to Hay (August 9, 1903), quoted in Pringle, *Theodore Roosevelt*, 198–199.

39. For the Ranger policing of Globe in October, see the *Report of the Governor of Arizona, 1903*, "Arizona Rangers." Captain Rynning commented that the Rang-

ers protected property and kept the peace. "We succeeded in settling the trouble amicably," he reported. Also see the Tucson *Citizen* (October 24, 1902).

40. Ranger Arrest Records for 1902; Ranger Financial Records for 1902; personnel record of McDonald Robinson.

### 1903: Peacekeeping at Mines and Saloons

1. Ranger personnel records. MacDonald, a native of the island of Mauritius, off the east coast of Africa, worked as a western cowboy before joining the Rangers.

2. Rynning, *Gun Notches*, 207.

3. The shooting in the Cowboy's Home Saloon is described in the Tucson *Citizen* (February 12, 1903) and the Bisbee *Review* (February 14, 1903).

4. Bisbee *Review* (February 14, 1903).

5. *Ibid.*

6. Rynning, *Gun Notches*, 208–209.

7. *Ibid.*, 213–214.

8. Bisbee *Review* (February 14, 1903).

9. Ranger Webb's trial is described in the Bisbee *Review* (February 14, 1903) and Rynning, *Gun Notches*, 210. Regarding criticism of Webb's exoneration, some witnesses alleged that Webb and other Rangers had been carousing from one saloon to another, and a bartender claimed that Webb had fired a shot in the Cowboy's Home a few minutes before he killed Bass. Critics of the Rangers groused that the organization had used its influence to exonerate Webb and maintain the good name of the force. Bisbee *Review* (February 14, 1903) and *Arizona Silver Belt* (July 2, 1903).

10. Act No. 64 expanding the Ranger company is in *Acts, Resolutions and Memorials, Twenty-second Legislative Assembly, 1903*, 104–106.

11. George E. Virgines has studiously examined the subject of Ranger badges in "The Arizona Rangers," *Arms Gazette*, 28.

12. Quotation from Joe Pearce in Ernenwein, "Lucky Star," *Ranch Romances*, 44.

13. For Harry Wheeler's promotion to sergeant see Governor Alexander O. Brodie to Wheeler (October 15, 1903). Ranger personnel records detail the other promotions. Also see the *Border Vidette* (November 7, 1903) in Peck memoirs, and Rynning to Joe Pearce (n.d., 1903), quoted in Pearce, "Line Rider," 101–102.

14. Ranger personnel records.

15. *Ibid.*

16. *Ibid.*

17. Pearce, "Line Rider," 101.

18. *Ibid.*, 101–102.

19. For Pearce's enlistment experience see Pearce, "Line Rider," 102–105.

20. The warrants of authority issued to the Rangers are frequently mentioned in official correspondence. See, for example, Harry Wheeler to Sims Ely (January 9, 1908).

21. For Rynning's training and arrest procedures see Pearce, "Line Rider," 107–110, and Pearce and Summers, "Joe Pearce — Manhunter," *Journal of Arizona History*, 260.

22. Pearce, "Line Rider," 110.

23. *Border Vidette*, 366, in Peck memoirs.

24. Ranger preferences for wearing holsters and revolvers are described by

Pearce in "Line Rider," 109–110, and in Ernenwein, "Lucky Star," *Ranch Romances*, 44. Pearce's blackjack quote is on 168.

25. Mine wages of $2.50 to $3.50 are related in the *Arizona Blade and Florence Tribune* (May 30, 1903).

26. For an excellent study of the entire situation at Clifton and Morenci, see Park, "The 1903 'Mexican Affair' at Clifton," *Journal of Arizona History*, 119–148.

27. Miners' leadership (Lastaunau, Salcido, and Colombo) is detailed in the *Arizona Daily Star* (June 11, 1903).

28. For details of the strike see Colquhoun, *Clifton-Morenci Mining District*.

29. Bud Bassett's typed reminiscences, 1.

30. *Ibid.*

31. For the initial stage of the Ranger trek to Morenci see: Bassett, typescript, 1; Rynning, *Gun Notches*, 231; Bassett to Mulford Winsor (March 15, 1936).

32. Bassett, typescript, 1.

33. *Ibid.*, 2, 4.

34. The activities of the Rangers as strikers marched toward Morenci are related in: Bassett, typescript, 2; Bassett to Winsor (February 14, 1936); Arthur Hopkins to Winsor (April 28, 1936).

35. Clifton *Copper Era* (June 11, 1903).

36. The arrest of Lastaunau is described in Bassett, typescript, 3.

37. *Ibid.*, 4.

38. The congregation of the Arizona National Guard and the U.S. Cavalry is described in the Clifton *Copper Era* (June 11, 1903).

39. Information about the famous Ranger photos at Morenci is in Virgines, "The Arizona Rangers," *Arms Gazette*, 31.

40. It should be mentioned that Sheriff Jim Parks also was given "one of the most handsome watches that has ever been seen in Arizona" by the Detroit, Arizona, and Shannon copper companies. Tucson *Citizen* (December 14, 1903).

41. Lastaunau's troubled imprisonment is related in Jeffrey, *Adobe and Iron*, 78–79, 98–99, and the Tucson *Citizen* (September 1, 1906).

42. For a sample of the criticism of Rangers as strikebreakers see the *Arizona Blade and Florence Tribune* (June 20, 1903).

43. The exodus of Rangers following the strike may be noted in the personnel records.

44. Rynning's July and August activities are detailed in the Tucson *Citizen* (August 28 and November 25, 1903).

45. Ranger cooperation with the Livestock Sanitary Board and the Arizona Cattle Growers' Association is described frequently in Reports of the Governor on the Rangers. The ranchers' loan of horses to Rangers pursuing rustlers is mentioned in Pearce, "Line Rider," 111. The Foster-Pruett recovery of stolen cattle is described in the Tucson *Citizen* (December 14, 1903).

46. Ranger Bailey's courageous arrest of three rustlers is related in a letter written in 1937 from Joe Pearce to Mulford Winsor, and in the Tucson *Citizen* (December 24, 1903).

47. The later activities of rustler Farrel are described in a letter from Harry Wheeler to Governor Kibbey (August 19, 1908).

48. *Report of the Governor of Arizona, 1904*, "Arizona Rangers," 79.

49. Rynning's method of reporting arrests and the Ranger policing of Pearce is in Captain Rynning's Report to the Governor (July 1, 1903), quoted in *Report of the Governor of Arizona, 1903*, "Arizona Rangers."

50. Tombstone *Epitaph* (December 21, 1903).

## 1904: A Few Tarnished Badges

1. Tucson *Citizen* (April 16, 1904).

2. Brooks's forceful arrest of Douglas is described in the Tucson *Citizen* (September 2, 1904).

3. Wood related his harassment by Rangers in an interview in Nogales, printed in the *Border Vidette* (November 12, 1945). Wood was seventy-four in 1945; he died in 1962 at the age of ninety-one. See the *Border Vidette*, 360, in Peck memoirs.

4. "Reminiscences of Samuel J. Hayhurst," typescript, 2.

5. This account of the posse's raid on the Stiles-Alvord hideout was published in the *Border Vidette* (February 27, 1904). Other accounts relate that Brooks and Lewis waylaid the outlaws in Nigger Head Canyon. See, for example, Hayhurst, typescript, 3.

6. Arcus Reddoch described his encounter with Alvord in the *Border Vidette*, 358, in Peck memoirs.

7. The immediate fate of Alvord and Stiles was related in the Tucson *Citizen* (May 20, 1905).

8. More than thirty years later Rynning recalled that this incident occurred during the spring "in about 1903. . . ." [Newell Jones, interview with Rynning, *Tribune* (July 3, 1936)]. But Rynning was accompanied by Johnny Brooks, who did not enlist until October 10, 1903, so these arrests evidently took place in the spring of 1904. Such an apprehension was not recorded in 1903, and detailed arrest records for succeeding years no longer exist.

9. Rynning, *Gun Notches*, 249 and 252. Also see Egerton, "The Case of the Cow Thief," *Arizona Republic* (March 28, 1982).

10. Rynning related his unique plan of detecting the rustlers in *Gun Notches*, 249–252.

11. *Ibid.*, 251–252.

12. Sam Hayhurst's quote is in the Hayhurst typescript, 3.

13. *Report of the Governor of Arizona, 1904*, "Arizona Rangers," 79.

14. Pearce, "Line Rider," 111.

15. The Ranger pursuit into Mexico after stock thieves was depicted by Pearce in "Line Rider," 111–120. This incident occurred early in his tenure as a Ranger, probably in 1904. Pearce served from November 23, 1903, until July 21, 1905.

16. Pearce, "Line Rider," 118.

17. Joe Pearce told the story of his mission to the Verde River country in "Line Rider," 167–178.

18. Rynning's report is summarized in *Report of the Governor of Arizona, 1904*, "Arizona Rangers."

19. The thousand-mile ride of Rangers Beaty, Bailey, and Devilbiss is described by Arthur A. Hopkins in a letter to Mulford Winsor (April 28, 1936).

20. For Jeff Kidder's background see: Tucson *Citizen* (April 4, 1908); De-Arment, "Arizona Ranger Jeff Kidder," Tombstone *Epitaph*, 9–11.

21. Arcus Reddoch, in the *Border Vidette*, 345, 359, and 360, in Peck memoirs.

22. *Ibid.*, 360.

23. The Milton-Kidder clash is described by Milton's biographer, J. Evetts Haley, in *Jeff Milton*, 364–365, and by Arcus Reddoch in the *Border Vidette*, 360, Peck memoirs. Another contemporary, Billy Bower, reminisced (*Border Vidette*, in Peck, 361): "I saw Jeff Milton and Jeff Kidder shoot across the street. Jeff on the

saloon side and Milton behind a pole — Castlebaum's store. You could see chips knocked off the post." I have found no other evidence of the incident reported by Bower.

24. Kidder's search for Alvord and Stiles was related by Reddoch in the *Border Vidette*, 359, in Peek memoirs.

25. The pistol-whipping of Radebush was angrily reported in the Bisbee *Review* (July 6, 1904).

26. Douglas *American* (July 7, 1904).

27. The inconclusive resolution of the case may be followed in miscellaneous newspapers quoted in Miller, *Arizona Rangers*, 76–77.

28. The case against Henry Gray was reported in the Tucson *Citizen* (October 13, 18, 19, 1904).

29. I have collected a considerable amount of biographical information on Harry Wheeler, consisting of correspondence with Priscilla R. Smith, supervisor, Certification Unit, Office of Vital Statistics, State of Florida, and with S. Morgan Slaughter, clerk of the Circuit Court, Duval County, Florida; Col. William B. Wheeler, Military Personnel Record, National Records Center, St. Louis, Missouri; Allyn Wheeler's gravestone in the City Cemetery, Tombstone, Arizona (which provided the only source I have ever found for Harry *Cornwall* Wheeler's middle name — his mother's maiden name was Cornwall); letter from Kenneth W. Rapp, assistant archivist, United States Military Academy (February 11, 1985); Harry C. Wheeler, Military Personnel Record, National Records Center, St. Louis, Missouri; Wheeler's Ranger enlistment papers are in his personnel file in the State Archives in Phoenix.

30. Rynning's praise is in Wheeler's personnel report (July 6, 1905). Wheeler was informed of his promotion to sergeant in a letter from Governor Brodie (October 15, 1903).

31. The shooting in the Palace Saloon is described in the Tucson *Citizen* (October 21 and November 5, 1904).

32. For Wheeler's aborted pursuit, read the Tucson *Citizen* (November 1, 1904).

33. Tucson *Citizen* (October 21, 1904).

34. The shooting by Johnny Brooks is reported in the Tucson *Citizen* (September 2, 1904). The fatal shooting of John Robinson in Naco at the hands of an unnamed Ranger (who was exonerated) is an obscure incident. The fatality is mentioned in the *Report to the Governor of Arizona, 1904*, "Arizona Rangers," 78, and it may have occurred in the last half of 1903 or the first half of 1904. It is described in slightly greater detail in a discussion of one of Captain Rynning's reports in the *Border Vidette* (January 21, 1905).

35. The arrest of Nuñez is reported in the *Border Vidette* (December 10, 1904).

### 1905: Riding Into Danger

1. For details of the creation of the New Mexico Mounted Police see Hornung, "Fullerton's Rangers," Tombstone *Epitaph*, 7–9. Quotation ("identical in all but the name . . .") is from the Tucson *Citizen* (April 20, 1905).

2. Pearce told the story of his expedition with Kidder and Rountree in Chapter 15 of his "Line Rider" m.s., 139–154.

3. *Ibid.*, 153.

4. Tucson *Citizen* (February 11, 1905). The *Citizen* was ardently Democratic and disapproved of the Republican president. "We do not say that Mr. Roosevelt

could have done better; but he certainly could have done worse." The *Citizen* acknowledged Kibbey's qualifications and gave grudging approval: "There is no reason why he should not make an entirely acceptable Governor."

5. Ranger opponents claimed that expenditures were more than $5,000 monthly, but the Governor's Report reveals that in the fiscal year 1904–1905, a total of $33,254.46 was spent on the company — an average of $2,771.20 per month. The Tucson *Citizen* (June 29, 1905) reported that newspapers in both Bisbee and Douglas stated that the Ranger company was "double its necessary size." Since the creation of the force, Rangers had been held in low regard in Bisbee. And even though headquarters currently was in Douglas, the 1903 killing of Lon Bass by Pvt. William Webb seems to have damaged the Ranger reputation there. The Douglas *International*, however, frequently supported the Rangers: "No body of men has done more to rid this territory of notorious cattle thieves, thugs and highwaymen."

6. The 1,052 Ranger arrests of fiscal year 1904–1905 break down as follows:

261 felonies:

| | |
|---|---|
| 9 murder | 28 horse theft |
| 23 felonious assault | 13 escaped prisoners |
| 31 burglary | 65 miscellaneous felonies (smuggling, |
| 15 robbery | passing counterfeit money, desertion |
| 19 swindlers, embezzlement, forgery, etc. | from U.S. Army, violation of immi- |
| 23 grand larceny | gration laws, etc.) |
| 35 cattle theft | |

778 misdemeanors:

| | |
|---|---|
| 299 drunk and disorderly | 17 keeping and frequenting opium re- |
| 62 assault | sorts |
| 28 petit larceny | 6 conducting bunco games and gam- |
| 48 carrying concealed weapons | bling without license |
| 20 violation of butcher license and stock law | 298 other misdemeanors and vagrant |

7. Kibbey's praise of the Rangers is quoted from *Report of the Governor of Arizona, 1905,* "Arizona Rangers."

8. The Ranger arrest of Hobbs is detailed in the Tucson *Citizen* (April 10, 1905).

9. Criticism of the Rangers in Graham County, quoted in conjunction with Private Rountree's case, is from the Bisbee *Miner,* as quoted in the Tucson *Citizen* (June 22, 1905).

10. Tucson *Citizen* (May 22, 1905).

11. The capture of Rivera is reported in the Tucson *Citizen* (September 26 and 29, 1905).

12. Tucson *Citizen* (June 15, 1905).

13. For Brooks's posse into Mexico see the Douglas *Dispatch,* quoted in the Tucson *Citizen* (June 22 and July 19, 1905).

14. For the recovery of Riley's body, see Arthur Hopkins to Mulford Winsor (April 28, 1936) and the Tucson *Citizen* (October 12, 1905).

15. For the arrest of Rodriguez and Parra, see the Tucson *Citizen* (October 3, 4, 5, and 6, 1905).

16. *Border Vidette* (June 3, 1905).

17. Tucson *Citizen* (June 5, 1905).

18. Kibbey's appointment of Millay instead of Rynning is discussed in the Douglas *Dispatch,* 1905 issues quoted in Miller, *Arizona Rangers,* 99–101.

19. The circumstances of Brooks's resignation are discussed in the Tucson *Citizen* (July 10, 1905) and Rynning, *Gun Notches*, 284–286.

20. Ranger personnel records.

21. Tucson *Citizen* (July 14 and 28, 1905).

22. *Ibid.* (August 11, 1905).

23. Rynning's inspection trip is described in the Tucson *Citizen* (August 11 and September 20, 1905).

24. Miller, *Arizona Rangers*, 115–116.

25. Enlistment details are available in the personnel records.

26. *Ibid.*

27. McPhaul's background is given in his biographical file at the Arizona Heritage Center in Tucson.

28. McPhaul's arrest of Guero was reported in the Yuma *Sun* (September 15, 1905).

29. Quoted on page 4 of a typescript under McPhaul's biographical file, Arizona Heritage Center, Tucson.

30. Yuma *Sun* (October 20, 1905).

31. For Holmes's duel with Arviso see: *Arizona Republican* (November 2, 1905); Tucson *Citizen* (November 3, 1905); *Report of the Governor of Arizona, 1906*, "Arizona Rangers," 21.

32. For Holmes's duel with "Matze Ta 55" see: *Report of the Governor of Arizona, 1906*, 21; and Holmes's personnel record.

33. For the roughing-up of Baldwin, see the *Graham County Advocate*, quoted in the *Border Vidette* (April 6, 1907).

34. For Grindell's visit to Tiburon see the Tucson *Citizen* (June 10, 1904).

35. Colonel Kosterlitzky to Captain Rynning (October 16, 1905).

36. For Hoffman's ordeal see Winsor, "Arizona Rangers," *Our Sheriff and Police Journal*, 56.

37. Tucson *Citizen* (October 16, 1905).

38. For the activities of the party see the Tucson *Citizen* (November 10 and 13, 1905).

39. *Ibid.* (November 25 and December 22, 1905).

40. *Ibid.* (December 22, 1905).

41. The murder of Plunkett and Kennedy was reported in the Tucson *Citizen* (July 22, 1905).

42. The organization of a special Ranger pursuit posse is described in the Tucson *Citizen* (August 7, 1905). Also see the personnel files of Hickey and Shute.

43. Ranger personnel records.

44. The story of the rigorous search through Sonora was related by Wheeler in 1910 to a reporter for the Tucson *Star*, quoted in Miller, *Arizona Rangers*, 109–115.

## 1906: Maintaining a Stronghold

1. For the story of Wheeler versus the Douglas criminals see the Tucson *Citizen* (January 24, 1906).

2. For Wheeler's apprehension of Howard see the Tucson *Citizen* (January 29, 1906).

3. For Wheeler's clash with Jiminez, and the resulting correspondence between Rynning and Torres, see the Douglas *Dispatch*, a 1906 issue quoted in Miller, *Arizona Rangers*, 126–127.

4. For the big posse sweep into Papago country, see the Tucson *Citizen* (March 27, 1906).

5. Greene's development of Cananea is described in the Tucson *Citizen* (March 10 and April 24, 1902). The quote about "rolling stock" is in the March 10 issue. Greene is the subject of C. L. Sonnichsen's award-winning biography, *Colonel Greene and the Copper Skyrocket.*

6. For the demands of the workers and the beginning of the riots see the Douglas *Dispatch* (June 2, 1906) and the Tucson *Citizen* (June 2, 1906). Paul J. Vanderwood, in *Disorder and Progress*, 141–149, places the Cananea incident into perspective as a part of the unraveling of Porfirio Diaz's regime. Also see Brayer, "The Cananea Incident," *New Mexico Historical Review*, 387–415.

7. Tucson *Citizen* (June 4, 1906) and Rynning, *Gun Notches*, 293 and 306.

8. The pleas from Cananea are detailed in the Tucson *Citizen* (June 4 and 5, 1906).

9. The arrivals of the trains from Cananea are reported in the Douglas *Dispatch* (June 2, 1906).

10. Rynning, *Gun Notches*, 292. The drills "to get the feel of discipline into them" are described in *Gun Notches*, 297.

11. *Ibid.*, 297. See also Hayhurst in his typed "Reminiscences," 2.

12. Bisbee *Review* (June 2, 1906).

13. For the fighting near Naco see the Tucson *Citizen* (June 2, 1906).

14. Rynning, *Gun Notches*, 299.

15. Tucson *Citizen* (June 2, 1906).

16. The arrival in Cananea and early activities of Rynning and his force are related in *Gun Notches*, 302, and the Tucson *Citizen* (June 2, 1906).

17. Hayhurst, quoted in his typed "Reminiscences," 2.

18. Rynning told of his skirmish near the hospital in *Gun Notches*, 304–305.

19. Kosterlitzky's arrival and the activities that night are reported in the Tucson *Citizen* (June 5, 1906) and in Liggitt, *My Seventy-Five Years Along the Mexican Border*, 68–72. Liggitt was a resident of Cananea and his account is quite informative.

20. The winding-down of the crisis is described in the Tucson *Citizen* (June 4 and 5, 1906) and Rynning, *Gun Notches*, 310–311.

21. Hayhurst, quoted in his typed "Reminiscences," 4.

22. Enlistments and discharges for 1906 are detailed in the personnel records. Burnett's discharge came after he got drunk in Tucson; when his behavior was reported to Rynning, the captain investigated, then revoked Burnett's commission. Tucson *Citizen* (July 14, 1906).

23. For Hopkins's appointment as undersheriff see the Tucson *Citizen* (November 16, 1906) and the letterhead on Cochise County stationery of the day.

24. Eperson's hearing and exoneration are reported in the *Border Vidette* (August 18, 1906) and the Tucson *Citizen* (August 6 and 10, 1906).

25. Tucson *Citizen* (August 10, 1906).

26. The Tucson *Star* interview with Wheeler is quoted in the *Border Vidette* (August 18, 1906).

27. Rynning's inspection of the San Carlos Reservation was reported in the Tucson *Citizen* (August 25, 1906).

28. Rynning's address to the Arizona Cattlemen's Association was reported in the Tucson *Citizen* (November 15, 1906) and the *Arizona Republican* (November 16, 1906).

29. As quoted in the *Arizona Republican* (November 16, 1906), the resolution read in full:

The Association fully appreciates the good work done by the Ranger force within our territory. Crime is less frequent within our territory than ever before, life and property are now almost absolutely secure, and the professional criminals have been driven from our territory and we are now enjoying a greater freedom from crime than any state in the union, and we petition the next session of our legislature and our governor not to reduce the Ranger force at the present time.

30. The Helvetia robbery and Stanford's pursuit was reported in the Tucson *Citizen* (June 2, 1906).

31. *Ibid.* (August 17, 1906).

32. Kibbey's border alert to the Rangers is related in the *Report of the Governor of Arizona, 1906,* "Arizona Rangers," 22.

33. The Ranger raid against revolutionaries was reported by Arthur A. Hopkins in a letter to Mulford Winsor (April 28, 1936) and in the Tucson *Citizen* (May 15, 1908).

34. The efforts of Old, Clarke, and Murphy against anarchists was reported in the Tucson *Citizen* (September 3 and 5, 1906).

35. *Ibid.* (September 6, 1906).

36. *Ibid.* (November 8 and 23, 1906).

37. The activities of Rangers against arms sales to Yaquis are described in the Tucson *Citizen* (February 25, 1907) and the *Report of the Governor of Arizona, 1906,* "Arizona Rangers," 21–22.

38. The efforts of Kidder and Rankin were described in the Tucson *Citizen* (May 1, 1908).

39. The quotations about Kidder's marksmanship are in the Tucson *Citizen* (April 6, 1908). Kidder had his name engraved on the backstrap of his fancy Colt. He sent this gun (serial number 246844) back to the factory late in 1907 and it was returned in mid-January 1908. During the interim, Kidder carried a single-action Colt .45 with a four-and-three-quarter-inch barrel. Late in her life Jeff's mother *gave* one of his Colts to a collector in San Jacinto, California. See Donoho, "Death of an Arizona Ranger."

40. The Kidder-Sparks scuffle against hobos was reported in a 1906 issue of the Bisbee *Review,* quoted in Miller, *Arizona Rangers,* 117.

41. Kidder's temporary transfer to Douglas late in 1906 was reported in the *Border Vidette* (December 8, 1906).

42. Kidder's shooting of Woods, and the subsequent hearing, were described in the Bisbee *Review* (December 31, 1906).

43. *Ibid.* (January 10, 1907).

## 1907: A Tighter Grip on the Force

1. The background of Silverton, Tracy, and their paramour was reported in the Bisbee *Review* (March 1, 1907) and the Tucson *Citizen* (March 2 and 4, 1907). Reddoch's comments may be found in the *Border Vidette,* 363, in Peck memoirs. These three sources also provide the best description of the Tracy-Wheeler duel. In addition see: Rynning, *Gun Notches,* 278–282; Winsor, "The Arizona Rangers," *Our Sheriff and Police Journal,* 58–59; Tucson *Citizen* (February 28, 1907).

2. Bisbee *Review* (March 1, 1907).

3. *Ibid.*

4. Tucson *Citizen* (March 4, 1907).

5. Wheeler's recuperation and refusal of the reward was reported in the Tucson *Citizen* (June 25, 1907) and in Liggitt, *My Seventy-Five Years Along the Mexican Border*, 53.

6. The arrest by Hayhurst and Rollins was reported in the Tucson *Citizen* (March 27, 1907).

7. Millay's resignation was reported in the Tucson *Citizen* (February 19, 1907). Rynning's appointment and arrival in Yuma were reported in Rynning, *Gun Notches*, 320, and the *Border Vidette* (March 30, 1907).

8. Sheriff White to Governor Kibbey (January 23, 1907).

9. Predictions of Wheeler's promotion may be found in the Tucson *Citizen* (February 19 and 28, 1907). Wheeler's promotion is confirmed in his Ranger personnel file.

10. The improving moral climate of Douglas was remarked upon in the Tucson *Citizen* (December 2, 1905).

11. Interesting descriptions of Naco are in the *Citizen* (September 26, 1901; December 24, 1902; June 12, 1906); Liggitt, *My Seventy-Five Years Along the Mexican Border*, 25; and Peterson, "Naco," *American West*. Wheeler's baseball challenge was reported in the *Citizen* (August 10, 1907).

12. General Orders (June 1, 1907), Ranger Records, Arizona State Archives, Phoenix. In a June 2 letter to the governor, Wheeler stated: "I have the honor to enclose, for your consideration, General Orders one to seven, of this year." But only six orders are pinned to this letter. "General Order #6," however, was originally typed #7, suggesting that Wheeler simply miscounted before the strikeover was made.

13. For Wheeler's scrupulous observance of border protocol see his letter to Sims Ely (June 8, 1907).

14. Wheeler to Governor Kibbey (June 2, 1907).

15. *Ibid.*

16. *Ibid.* (August 15, 1907).

17. Enlistments and discharges under Wheeler are detailed in the personnel records.

18. The Emett controversy may be followed in: [Illegible] to B. F. Saunders (December 1, 1907); [Illegible] to E. S. Clark (December 5, 1907); B. F. Saunders to Judge John J. Hawkins (December 9, 1907). These letters are in the Ranger correspondence in the Arizona State Archives.

19. Wheeler to Sims Ely (October 24, 1907).

20. *Ibid.*

21. The arrest records and personnel records prove Larn's disappointing performance. Larn dated his letter of resignation December 3, 1907, but this date was scratched out and someone, in a clearly different style of handwriting, wrote in "January 4, 1908." Larn's personnel file records his resignation as December 21, 1907. Larn's letter, addressed to "Cap C H Wheeler," reads:

> Pleas Except my resignation as Territorial Ranger. I will make out my
> Report for Dec. 1907 as soon as Mr W A Olds returns from Clear Creek
> > Yours very Truly
> > W A Larn

On January 11 Wheeler forwarded Larn's resignation to the governor, commenting: "I never saw this man Larn, he was enlisted by Lieut. Old." Old apparently was deeply disappointed in Larn's performance; Wheeler gathered that the erstwhile undercover man "is not deserving of a good discharge." Wheeler to Governor Kibbey (January 11, 1908).

22. For the Olney-Wheeler confrontation, see: Olney to Wheeler (November 18, 1907); Wheeler to Governor Kibbey (November 26, 1907); Olney's personnel file.

23. McDonald to Wheeler (November 23, 1907).

24. Wheeler to Governor Kibbey (November 26 and 27, 1907).

25. Wheeler's personnel file.

26. Wheeler to Sims Ely (May 19, 1907).

27. *Ibid.*

28. *Border Vidette* (November 23, 1907).

29. Kidder's arrest of Larrieau was referred to in Wheeler to Kidder (December 18, 1907).

30. Ranger performance in the latter months of 1907 may be studied in the monthly reports for August, October, November, and December.

31. Wheeler's desire to station his men in pairs was related in his Monthly Report (July 1908). Beaty's corroborating remark is quoted in the *Border Vidette*, 365, in Peck memoirs.

32. Wheeler requested a machine gun in letters to Governor Kibbey (July 18 and 27, 1907).

33. The 1907 attempts to abolish or reduce the Ranger company may be followed in the Tucson *Citizen* (February 26 and March 4, 1907). Kibbey is quoted in *Report of the Governor of Arizona, 1907,* "Arizona Rangers, 13–14.

34. The act preventing "Steer Tying Contests" was described in the Tucson *Citizen* (April 3, 1907).

35. Bisbee *Review* (April 14, 1907).

36. *Ibid.*

37. Criticism against gamblers ("men of the green table") quoted from the Tucson *Citizen* (November 23, 1904).

38. Descriptions of the Nogales resistance to the anti-gambling laws are described in letters from Kidder to Wheeler (May 16, 1907) and from the Arizona attorney general to Kidder (June 14, 1907).

39. Captain Wheeler discussed the situation in two letters to Governor Kibbey (August 27 and 29, 1907).

40. The problem in Yuma was described in: Frank Wheeler to Harry Wheeler (November 23, 1907); and Peter Robertson to Governor Kibbey (September 7, 1907). For resistance elsewhere in the territory to anti-gambling efforts, see: Harry Wheeler to Governor Kibbey (November 25, 1907); Wheeler's monthly reports for October, November, and December, 1907.

41. The incident involving the Douglas circus parade was described by Joe Pearce in his "Line Rider" m.s., 119–120.

42. For detailed information on Mother Jones, see: Jones, *Autobiography of Mother Jones;* Atkinson, *Mother Jones, The Most Dangerous Woman in America;* Fetherling, *Mother Jones, The Miners' Angel.*

43. Mother Jones's arrival in Arizona was described in the Tucson *Citizen* (February 22, 1907).

44. For the story of the Sarabia kidnaping, see: Tucson *Citizen* (July 3, 1907); Bisbee *Review,* 1907 issue quoted in Miller, *Arizona Rangers,* 147–154.

45. Jones, *Autobiography,* 139.

46. Sarabia's fate is described in Miller, *Arizona Rangers,* 155–156.

47. Tucson *Citizen* (April 29, 1907).

48. Stations for Old and his men may best be followed by inspecting the letterhead on stationery for the Northern Detachment. The monthly reports also provide details.

49. Tucson *Citizen* (August 5, 1907).

50. Frank Wheeler's long career as a Ranger may be traced in his personnel file.

51. Kerrick's early activities were reported in the Tucson *Citizen* (July 1, 1907).

52. Phoenix *Democrat,* 1907 issue quoted in Miller, *Arizona Rangers,* 160–161.

53. The fight is described in the Tucson *Citizen* (August 13 and 14, 1907).

54. For the aftermath of the fight, see the Tucson *Citizen* (July 31 and August 13, 1907).

55. The ominous situation in El Cubo was described in the Tucson *Citizen* (September 16, 1907).

56. The formation and progress of the posse was reported by Captain Wheeler in a letter to Governor Kibbey (September 14, 1907). Also see Miller, *Arizona Rangers,* 163–170.

## 1908: Loss of a Fighter

1. For the efforts of Gunner and Kidder against gunrunners early in 1908, see: Wheeler to Governor Kibbey (January 9 and 18, 1908); Gunner to Wheeler (January 31, 1908).

2. The gambling problem is described in: Wheeler to Governor Kibbey (January 9 and 22, February 1, 1908); Gunner to Wheeler (January 31, 1908). For the implication of Constable Ells, see Wheeler to Frank Cleaveland (June 9, 1908).

3. Wheeler described his attempt to halt the roulette game in Naco in two letters to Governor Kibbey (February 3 and 7, 1908). Wheeler's confrontation with the saloon owners was reported in the Tucson *Citizen* (February 13, 1908).

4. Wheeler's activities, and Hayhurst's trip to Texas, were reported in Wheeler to Governor Kibbey (January 9 and 22, 1908).

5. Wheeler's stopover in Yuma was reported in Wheeler to Governor Kibbey (January 31, 1908).

6. Wheeler to Governor Kibbey (January 14, 1908).

7. For Gunner's undercover activities see: Gunner to Wheeler (February 29, 1908); Peter Robertson to Wheeler (February 21, 1908); Wheeler to Sims Ely (February 23, 1908).

8. Gunner to Wheeler (February 29, 1908); and Robertson to Wheeler (February 28, 1908).

9. Gunner to Wheeler (March 3, 1908).

10. *Ibid.*

11. *Ibid.* (March 5, 1908).

12. The dismayed complaints of Gunner and Robertson are in: Gunner to Wheeler (March 5, 1908) and Robertson to Wheeler (March 5, 1908).

13. Gunner to Wheeler (March 6, 1908).

14. Wheeler to Governor Kibbey (March 7, 1908).

15. *Ibid.* (March 16, 1908).

16. *Border Vidette* (December 8, 1906).

17. *Ibid.* (February 15, 1908).

18. For Kidder's action against the "undesirables," see the *Border Vidette* (March 28, 1908).

19. For the recent complaints against Kidder, see: Willson, "Sergeant Jeff Kidder," *Arizona Republic;* and Wheeler to Governor Kibbey (April 8, 1908).

20. Willson, "Sergeant Jeff Kidder," *Arizona Republic.*

21. The account of the shooting and Kidder's final hours are taken from: Bisbee *Review* (April 4, 1908); Tucson *Citizen* (April 4 and 6, 1908); Wheeler to Governor Kibbey (April 8, 1908). Also see Donoho, "Death of an Arizona Ranger."

22. Bisbee *Review* (April 4, 1908).

23. For speculation and motivation about Kidder having been trapped, see the Tucson *Citizen* (April 6, 8, and 24, and May 1, 1908).

24. The disposal and condition of Kidder's body were related in: Bisbee *Review* (April 5, 1908); Tucson *Citizen* (April 6, 1908); Wheeler to Governor Kibbey (April 8, 1908).

25. Wheeler to Governor Kibbey (April 8, 1908).

26. For the sad ride of Wheeler and his men into Bisbee, see: Wheeler to Governor Kibbey (April 8, 1908); and Bisbee *Review* (April 7, 1908).

27. Bisbee *Review* (April 8, 1908).

28. Kidder's funeral services were described in the Tucson *Citizen* (April 8, 1908). For the plight of Jip see the Bisbee *Review* (April 5, 1908), and the Tucson *Citizen* (April 30, 1908).

29. Bisbee *Review* (April 4, 1908).

30. For the recovery of Kidder's revolver and badge, see: Wheeler to Governor Kibbey (April 10, 13, and 15, 1908).

31. Tucson *Citizen* (April 24, 1908).

32. *Ibid.*

33. Wheeler to Governor Kibbey (April 9, 1908).

34. *Ibid.* (April 12, 1908).

35. *Ibid.* (May 7, 1908).

36. This account of the Arnett shooting is taken from Wheeler's seven-page report of the incident — Wheeler to Governor Kibbey (May 7, 1908) — and from the Tucson *Citizen* (May 6, 1908) and the Bisbee *Review* (May 7, 1908). Also see Egerton, "A Brazen Horse Thief," *Arizona Republic* (July 12, 1981).

37. Bisbee *Review* (May 7, 1908).

38. For Heflin as an informer, see Wheeler to Acting Governor John H. Page (May 11, 1908).

39. A copy of Heflin's agreement with Wheeler is in the Ranger correspondence.

40. Heflin's revelation that Ells aided Arnett is in Wheeler to J. F. Cleaveland (June 10, 1908). Heflin's disclosure about the high Mexican official is in Wheeler to J. F. Cleaveland (May 15, 1908). For the exodus of wary badmen, see Wheeler to Cleaveland (May 16 and 19, 1908).

41. Wheeler to J. F. Cleaveland (May 19 and 29, 1908) and Wheeler to Governor Kibbey (May 26, 1908).

42. Wheeler to J. F. Cleaveland (June 4 and 6, 1908).

43. For the organization of Wheeler's posse, the enlistment of Humm, the leaves of absence, and the trek into Mexico, see: Wheeler to J. F. Cleaveland (June 4, 9, and 10, 1908) and Humm personnel file.

44. Wheeler to J. F. Cleaveland (June 19, 1908).

45. *Ibid.*

46. Sherlock-Bly's background and arrest were reported in the Tucson *Citizen* (July 21, 1908).

47. His pardon was reported in the Tucson *Citizen* (August 18, 1908).

48. Bisbee *Review* (August 5, 1908).

49. Bisbee *Review* (August 5, 1908) and Tucson *Citizen* (August 5, 1908).

50. The sad story of Downing's wife was related in the Tucson *Citizen* (August 5, 1908).

51. The location of the Free and Easy Saloon is given in Schultz, *Southwestern Town*, 42. Downing's troubles with Speed and Snow are described in the Bisbee *Review* (May 7, 1908) and Tucson *Citizen* (August 5, 1908).

52. Wheeler to Sims Ely (February 5, 1908) and Wheeler to Governor Kibbey (February 2, 1908). Speed's reenlistment and demotion are verified in his personnel file.

53. General Order #5 (June 1, 1907).

54. Wheeler quoted his instructions to Speed in a letter to Governor Kibbey (August 7, 1908).

55. Wheeler to Governor Kibbey (August 7, 1908) and Tucson *Citizen* (August 5, 1908).

56. The meeting of citizens with Wheeler in Benson was reported in Wheeler to Governor Kibbey (August 5, 1908).

57. Testimony of George McKittrick, Proceedings of the Coroner's Jury (August 5, 1908).

58. Testimony of Bud Snow, *ibid.*

59. Testimony of R. E. Cushman, *ibid.*

60. For details of the shooting, see: Bisbee *Review* (August 5, 1908); Tucson *Citizen* (August 5, 1908); and Proceedings of the Coroner's Jury (August 5, 1908).

61. Coroner's Verdict (August 5, 1908).

62. Wheeler's telegram was wired to Governor Kibbey on August 5. The quote from Wheeler about Downing is from a detailed report, Wheeler to Governor Kibbey (August 7, 1908).

63. Parmer's shooting of Van Valer was described in: Wheeler to Governor Kibbey (November 23, 1908) and Tucson *Citizen* (November 19 and 20, 1908).

64. The escape of Davlandos from Wheeler was reported in: Wheeler to Governor Kibbey (December 8, 1908); Wheeler to J. F. Cleaveland (December 10, 1908); and Monthly Report (November 1908).

65. Wheeler's discussion of the likely move of various Rangers to other offices was in Wheeler to Sims Ely (February 23, 1908).

66. Wheeler to Sims Ely (February 23, 1908).

67. Wheeler to Governor Kibbey (June 30, 1908).

68. Wheeler to J. F. Cleaveland (August 25, 1908).

69. For the outbreak of criminal activities in Naco in the absence of Rangers, see Wheeler to Governor Kibbey (October 16, 1908).

70. For Ranger activities in Williams, Octave, and Courtland, see, respectively, the Monthly Report for March, June, and December, 1908.

71. For Mickey's situation at Laguna Dam, see Wheeler to Governor Kibbey (June 30 and July 30, 1908).

72. Wheeler's soul-searching in regard to Chase, see, in order: Wheeler to J. F. Cleaveland (November 2, 1908); Wheeler to Governor Kibbey (October 30, 1908); Wheeler to Governor Kibbey (October 28, 1908); Wheeler to Governor Kibbey (November 4, 1908).

73. Wheeler's disillusionment with Gunner was recorded in: Wheeler to Governor Kibbey (November 30, 1908); and Gunner's personnel file. In Gunner's fitness report, Wheeler stated that he was "a good man when not drinking."

74. For Wheeler's reaction to Black's arrest of his brother-in-law, see Wheeler to Governor Kibbey (August 19, 1908). For Wheeler's decision to discharge Black, see Wheeler to Governor Kibbey (November 30, 1908).

75. See A. E. Ehle's personnel file. For Anderson's resignation see Wheeler to Governor Kibbey (May 7, 1908).

76. The termination of Fraser is reported in Wheeler to Governor Kibbey (July 10, 1908). For Wheeler's conflict with Bates, see: J. F. Cleaveland to Wheeler (August 5, 1908) and Bates's personnel file.

77. Wheeler's anti-Texan feelings are described in letters to: Sims Ely (January 11, 1908); Governor Kibbey (February 2, 1908); and Sims Ely (February 5, 1908).

78. The peculiar circumstances of Horne's termination may be studied in: Wheeler to Governor Kibbey (April 27, 1908); Horne's discharge papers; and Wheeler to J. F. Cleaveland (April 29, 1908).

79. Wheeler's request for Springfields is outlined in a letter to Governor Kibbey (October 1, 1908). His request for bloodhounds is in the Monthly Report (October 1908), 2–3.

80. Wheeler's pleas to be allowed to make extensive inspection trips were made in two letters to Governor Kibbey (August 19 and October 1, 1908).

81. Wheeler's October sweep was described in his report to Governor Kibbey (October 25, 1908). Other forays are frequently mentioned in correspondence and newspapers.

82. Wheeler to J. F. Cleaveland (August 22, 1908).

83. The October capture of a horse thief was reported in Wheeler to Governor Kibbey (October 25, 1908).

84. Wheeler to Governor Kibbey (August 7, 1908).

85. Monthly Report (August 1908).

86. Wheeler to Governor Kibbey (August 4, 1908).

## 1909: Political Demise

1. For Wheeler's 1908 request to go to Florida, see: Telegram, Wheeler to Governor Kibbey (April 15, 1908); Wheeler to Governor Kibbey (April 15, 1908). For Wheeler's decision not to go, see Wheeler to J. F. Cleaveland (September 4, 1908). For the death of Major Wheeler, see the Tucson *Citizen* (December 3, 1908). For his 1909 request, see: Wheeler to Governor Kibbey (January 1, 1909); Telegram, Governor Kibbey to Wheeler (January 2, 1909).

2. Wheeler to Governor Kibbey (January 3, 1909).

3. For Wheeler's activities prior to departure, see Wheeler to Governor Kibbey (January 3, 1909).

4. For the dismissal of McGee and King, see: Wheeler to Governor Kibbey (January 25 and 26, 1909); personnel file, McGee; personnel file, King.

5. For Wheeler's inquiry about the letters of support, see Wheeler to J. F. Cleaveland (January 26, 1909).

6. Telegram, Acting Governor John H. Page to Wheeler (January 27, 1909).

7. For the situation in Globe, see: Telegram, Wheeler to J. F. Cleaveland (January 28, 1909); Telegram, Wheeler to J. F. Cleaveland (January 29, 1909);

Telegram, J. F. Cleaveland to Wheeler (January 29, 1909); Telegram, Wheeler to J. F. Cleaveland (January 30, 1909).

8. Tom Gadberry's personnel file.

9. For the Rangers as a campaign issue, see: Tucson *Citizen* (January 14, 1909); Winsor, "Arizona Rangers," *Our Sheriff and Police Journal,* 61.

10. Tucson *Citizen* (November 13, 1908).

11. *Ibid.* (November 19, 1908).

12. *Ibid.* (January 13, 1909).

13. For the Democratic caucus, see: *Arizona Republican* (February 16, 1909); Winsor, "Arizona Rangers," *Our Sheriff and Police Journal,* 61. For first-day activities of the Assembly, see the Tucson *Citizen* (January 18, 1909).

14. The full text of the governor's Tuesday message is given in the Tucson *Citizen* (January 19, 1909).

15. For Weedin's introduction of the bills to abolish the Rangers and the office of public examiner, see the Tucson *Citizen* (January 23, 1909). For the Deputy Ranger bill, see the *Citizen* (January 16, 1909).

16. For an example of Democratic criticism of the Rangers, see the *Arizona Republican* (February 16, 1909).

17. Tucson *Citizen* (January 25, 1909).

18. Winsor, "Arizona Rangers," *Our Sheriff and Police Journal,* 61.

19. Tucson *Citizen* (January 29, 1909).

20. *Ibid.* (January 30, 1909).

21. Wheeler described the decline in Ranger morale and the perceived increase in crime in two letters to Governor Kibbey (February 6 and 10, 1909). For the letter from his headquarters clerk, see Emil Lenz to Acting Governor John H. Page (February 3, 1909).

22. Wheeler to Governor Kibbey (February 5, 1909).

23. For Wheeler's offer to resign, see: Wheeler to Governor Kibbey (February 5, 1909); Tucson *Citizen* (February 8, 1909). For Wheeler's invitation to Phoenix, see Telegram, J. F. Cleaveland to Wheeler (February 11, 1909).

24. The Tucson Chamber of Commerce plea was reported in the Tucson *Citizen* (February 12, 1909). For samples of other support for the company, see the Ranger correspondence for February 1909; "do all in your power . . .," quoted from Marshall Young to Governor Kibbey (February 9, 1909).

25. Slaughter's statement of support was run in the Tucson *Citizen* (February 3, 1909). Mossman's plea was in the *Citizen* (February 8, 1909), and Wheeler's appeal was in the *Citizen* (February 11, 1908).

26. Finley's letter to the editor was published in the Tucson *Citizen* (February 12, 1909). The editorial reply came in the *Citizen* (February 13, 1909).

27. For hope that the caucus might relent, see the Tucson *Citizen* (February 15, 1909).

28. The return of the bills to each house, the text of the governor's veto message, and the subsequent legislative action is described in: *Arizona Republican* (February 16, 1909); Tucson *Citizen* (February 15, 1909). The official text of the governor's message is in *Legislative Journals, Territory of Arizona, 1909,* 94–121.

29. *Arizona Republican* (February 16, 1909).

30. Ranger expenses were declared in *Report of the Governor of Arizona, 1908,* "Arizona Rangers," 23.

31. *Arizona Republican* (February 16, 1909).

32. Section 3 of the abolishment bill is quoted from *Acts, Resolutions and Memorials, Twenty-fifth Legislative Assembly, 1909,* Chapter 4:3.

33. Ranger arrests on the last day were reported in Wheeler's Monthly Report (February 1909).

34. Bisbee *Review,* quoted in Miller, *Arizona Rangers,* 225.

35. Wheeler to Governor Kibbey (both letters are dated February 17, 1909).

36. The seventeen-page handwritten January report is in the Ranger correspondence. Wheeler mentions his Oliver typewriter in a letter to Governor Kibbey (February 19, 1909).

37. Wheeler's letter is quoted in Miller, *Arizona Rangers,* 223–225; G. L. Coffee (February 18, 1909) and J. Frank Wooton (February 19, 1909) were among those who wrote to Governor Kibbey.

38. Monthly Report (February 1909).

39. For Wheeler's discharge and the accompanying letter, see Governor Kibbey to Wheeler (March 25, 1909).

## Epilogue

1. For Wheeler's activities after the Rangers were abolished, I read the Tombstone *Prospector* (courtesy of the Arizona State Archives through interlibrary loan) covering 1909 through 1924. I have a substantial file of photocopied clippings which detail his professional activities as well as his personal life.

2. Any study of the Bisbee Deportation must begin with James W. Byrkit, *Forging the Copper Collar.* Arizona newspapers of the day, of course, were dominated by this story.

3. Wheeler's World War I record is available through the National Records Center, St. Louis, Missouri, and his efforts to enlist were avidly reported by the Tombstone *Prospector.*

4. I followed the story of Allyn Wheeler's accident and long illness via the Tombstone *Prospector.* For the divorce see: *Mamie O. Wheeler v. Harry C. Wheeler* (October 14, 1919), Superior Court Records, Cochise County Courthouse, Bisbee, Arizona; and the Tombstone *Prospector* (July 2 and 29, 1919). Wheeler's 1919 marriage license is on file at the Office of the County Clerk in El Paso. In 1980 his surviving daughter—Jessie Jacqueline Wheeler, I assume, although as of this writing I have been unable to locate her— visited the old Tombstone Courthouse and jotted down information about the offspring of Wheeler's second marriage. I have dozens of newspaper clippings on file which mention Wheeler's activities in rifle competitions.

5. Bisbee *Daily Review* (December 16 and 17, 1925).

6. For Mossman's later life see: Hunt, *Cap Mossman,* 219–240; *Arizona Daily Star* (January 9, 1947, and September 6, 1956); Roswell *Daily Record* (September 6, 1956).

7. For Rynning's later life see: Phoenix *Tribune Sun* (June 19, 1941); unlabeled obituary, Rynning file, Arizona Heritage Center, Tucson.

8. For Rye Miles see: Coolidge *News* (April 24, 1942); *Arizona Daily Star* (December 10, 1936).

9. For Joe Pearce see: *Apache County Independent News* (June 3, 1949); Tucson *Daily Citizen* (March 6, 1958).

10. For Bud Bassett see: "Honor the Past . . . Mold the Future," *Gila Centennial.*

11. For Oscar Felton see: Tucson *Citizen* (November 2, 1907).

12. For Luke Short see: Tombstone *Prospector* (May 21, 1914).

13. For Reuben Neill see: *Our Sheriff and Police Journal* (Vol. 28, No. 10).

14. For W. D. Allison see: *Arizona Daily Star* (April 5, 1923).

15. For Dayton Graham see: Tucson *Citizen* (July 8, 1903, and June 21, 1905) and Arcus Reddoch in the *Border Vidette,* 364, Peck memoirs.

16. For Billy Old see: Tucson *Citizen* (May 5, 1982).

17. For Burt Alvord see: Sonnichsen, *Billy King's Tombstone,* 92; Erwin, *The Southwest of John H. Slaughter,* 248; Coolidge, *Fighting Men of the West,* 190.

18. For Billy Stiles see: Tucson *Citizen* (December 21, 1908).

19. For the $100 monthly pension see: Phoenix *Gazette* (November 11, 1964).

20. For *26 Men* see: Parish and Pitts, *The Great Western Pictures,* 427.

21. For *Arizona Ranger* see: Hardy, *The Western,* 165.

22. For Chapo Beaty see: Phoenix *Gazette* (May 21, 1982); *Arizona Daily Star* (November 8, 1964).

23. *Los Angeles Times* (May 25, 1958) and *Arizona Daily Star* (March 15, 1963).

24. Tucson *Citizen* (May 5, 1982).

25. For the Winchester as controversy see: Phoenix *Gazette* (May 18, 1963).

26. The apology ceremony, illustrated with sequence photos, is described in the *Arizona Journal* (September 18, 1963).

# Bibliography

### Documents

Arizona Ranger Files at the Arizona State Archives, Phoenix. The Archives has collected the Ranger material in four boxes: *Box One* — Enlistment and discharge papers; *Box Two* — Miscellaneous correspondence; *Box Three* — Salary claims; *Box Four* — Expenses, receipts, and other financial data. *Note:* Ranger correspondence was stored in the basement of the old Capitol. On August 21, 1921, Cave Creek flooded west Phoenix, completely cutting off the Capitol and filling the basement with water. All documents were a soggy mass, and within a week reams of various papers had been spread out to dry on the Capitol grounds. High winds came up, and among the documents that blew away were the first six years of Ranger correspondence. Thirty-six letters survive from 1907; more than 160 from 1908; and about fifty from 1909. A majority of the letters are from Capt. Harry Wheeler to Governor Joseph Kibbey or his secretary. Many of the other letters are from the governor's office to Wheeler, while there is a scattering of miscellaneous correspondence. Also included are sixteen monthly reports from 1907 to 1909.

The Arizona Historical Society, Arizona Heritage Center, Tucson, maintains a mine of information about the Rangers. A large Ranger file has been collected, biographical information is available on many Rangers, and a great variety of related materials are on deposit.

*Acts, Resolutions, and Memorials, Twenty-second Legislative Assembly, 1903.* Act 57, p. 93.

*Acts, Resolutions, and Memorials, Twenty-second Legislative Assembly, 1903.* Act 64, secs. 1–10, pp. 104–106.

*Acts, Resolutions, and Memorials, Twenty-fifth Legislative Assembly, 1909.* Act 4, p. 3.

General Orders (6) for the Arizona Rangers from Captain Harry Wheeler. June 1, 1907. Arizona Rangers Correspondence File, State Archives.

Governor's Message, "The Arizona Rangers," to the Twenty-fifth Legislative Assembly, 1909, pp. 22–23.

"Message of Governor Joseph H. Kibbey to the Council of the Twenty-fifth Legislative Assembly," *Journal of the Twenty-fifth Legislative Assembly, 1909,* pp. 94–111; reply of Councilman Thomas F. Weedin, pp. 111–122.

Report of Captain Burton Mossman of Arrests Made by Arizona Rangers from October 2, 1901 to June 30, 1902.

*Revised Statutes of the Arizona Territory, 1901.* Pars., 3213–30, pp. 833–836.

U.S. Dept. of the Interior, *Report of the Governor of Arizona to the Secretary of the Interior,* "Arizona Rangers," 1902, 1903, 1904, 1905, 1906, 1907, 1908, 1909.

## Manuscripts

Bassett, James H. Reminiscences. Typescript in Arizona Heritage Center, Tucson.

Pearce, Joe, and Richard Summers. "Line Rider." Lengthy typescript available at the Arizona Heritage Center, Tucson.

Peck, Arthur (Artisan) Leslie, Sr. Memoirs. Biographical Files, Arizona Heritage Center, Tucson. Lengthy typescript contains numerous excerpts from the Nogales *Border Vidette*.

"Reminiscences of Sam J. Hayhurst," as told to Mrs. George F. Kitt, September 27, 1937. Biographical Files, Arizona Heritage Center, Tucson.

## Letters

Miscellaneous correspondence available at the Arizona Heritage Center, Tucson:

Bassett, J. H., to Mulford Winsor (February 14, 1936, and March 15, 1936).

"C P C" to Harry Wheeler (September 23, 1925).

Donoho, Ron, to Lori Davisson (December 3, 1982).

Hopkins, Arthur A., to Mulford Winsor (April 28, 1936).

Letter from Joe Pearce (1937).

Pearson, Pollard, to Mulford Winsor (December 1936).

Rynning, Thomas, to Joe Pearce (October 1903).

## Interviews

Barnes, Will C., interviewed by Mulford Winsor (May 12, 1936).

Mossman, Burt, interviewed by Will C. Barnes at the Raleigh Hotel, Washington, DC (April 17, 1935).

Pace, Scott, interviewed by the author in Solomon, Arizona (March 12, 1983).

Payne, John, interviewed by the author in Naco, Arizona (March 12, 1983).

Zearing, Wally, interviewed by the author in Benson, Arizona (March 13, 1983).

## Newspapers

*Apache County Independent News*
*Arizona Blade and Florence Tribune*
*Arizona Daily Star*
*Arizona Journal Miner*
*Arizona Republic*
*Arizona Republican*
*Arizona Silver Belt*
Bisbee *Review*
Clifton *Copper Era*
Coolidge *News*
Courtland *Arizonian*
Douglas *American*
Douglas *Dispatch*
Douglas *International*
*Graham County Advocate*
*Los Angeles Times*
*Navajo Apache Independent*

Nogales *Border Vidette*
Phoenix *Democrat*
Phoenix *Gazette*
Phoenix *Tribune Sun*
Roswell *Daily Record*
Solomonville *Bulletin*
Tombstone *Epitaph*
Tombstone *Prospector*
Tucson *Citizen*
Tucson *Star*
Yuma *Sun*
Miscellaneous Obituary Files, Arizona Heritage Center, Tucson, Arizona

*Books*

Atkinson, Linda. *Mother Jones, The Most Dangerous Woman in America*. New York: Crown Publishers, Inc., 1978.

Bailey, Lynn R. *Bisbee: Queen of the Copper Camps*. Tucson: Westernlore Press, 1983.

Ball, Larry D. *The United States Marshals of New Mexico and Arizona Territories, 1846–1912*. Albuquerque: University of New Mexico Press, 1978.

Brent, William, and Milarde Brent. *The Hell Hole*. Yuma, AZ: Southwest Printers, 1962.

Burgess, Opie Rundle. *Bisbee Not So Long Ago*. San Antonio: Naylor Company, 1967.

Byrkit, James W. *Forging the Copper Collar, Arizona's Labor Management War of 1901–1921*. Tucson: University of Arizona Press, 1982.

Chisholm, Joe. *Brewery Gulch*. San Antonio: The Naylor Company, 1949.

Colquhoun, James. *The History of the Clifton-Morenci Mining District*. London: John Murray, 1924.

Coolidge, Dane. *Fighting Men of the West*. New York: E. P. Dutton & Co., Inc., 1932.

Cosulich, Bernice. *Tucson*. Tucson: Arizona Silhouettes, 1953.

Erwin, Allen A. *The Southwest of John H. Slaughter, 1841–1922*. Glendale, CA: Arthur H. Clark Company, 1965.

Fetherling, Dale. *Mother Jones, The Miners' Angel*. Carbondale and Edwardsville, IL: Southern Illinois University Press, 1974.

Fuchs, James R. *A History of Williams, Arizona, 1876–1951*. Tucson: University of Arizona Press, 1955.

Haley, J. Evetts. *Jeff Milton: A Good Man with a Gun*. Norman: University of Oklahoma Press, 1948.

Hardy, Phil. *The Western*. New York: William Morrow and Company, 1983.

Heatwole, Thelma. *Ghost Towns and Historical Haunts in Arizona*. Phoenix: Golden West Publishers, 1981.

Hunt, Frazier. *Cap Mossman: Last of the Great Cowmen*. New York: Hastings House, 1951.

Jeffrey, John Mason. *Adobe and Iron*. La Jolla, CA: Prospect Avenue Press, 1969.

Jones, Mother [Ann]. *Autobiography of Mother Jones*. Edited by Mary Field Parton. Chicago: Charles H. Kerr, 1925.

Kelly, George H., comp. *Legislative History: Arizona, 1864–1912*. Phoenix: Manufacturing Stationers, 1926.

King, Frank M. *Wranglin' the Past*. n.p., 1935.

Liggitt, Wm. (Bill), Sr. *My Seventy-Five Years Along the Mexican Border.* New York: Exposition Press, 1964.

Martin, Douglas D., ed. *Tombstone's Epitaph.* Albuquerque: University of New Mexico Press, 1958.

Miller, Joseph. *Arizona, The Grand Canyon State.* New York: Hastings House, 1966 ed.

————. *The Arizona Rangers.* New York: Hastings House, Publishers, Inc., 1972.

Morison, Elting E., ed. *The Letters of Theodore Roosevelt.* 8 vols. Cambridge, MA: Harvard University Press, 1951–54.

Parish, James Robert, and Michael R. Pitts. *The Great Western Pictures.* Metuchen, NJ: The Scarecrow Press, Inc., 1976.

Patton, James M. *History of Clifton.* Clifton, AZ: Greenlee County Chamber of Commerce, 1977.

Pringle, Henry F. *Theodore Roosevelt.* New York: Harcourt, Brace and Company, 1931.

Raine, William MacLeod. *Famous Sheriffs and Western Outlaws.* Garden City, NY: Garden City Publishing Company, Inc., 1903.

Reps, John W. *Cities of the American West: A History of Frontier Planning.* Princeton, NJ: Princeton University Press, 1979.

Roosevelt, Theodore. *The Rough Riders.* New York: Signet Classic Edition, 1961 [1899].

Rynning, Thomas H. *Gun Notches, The Life Story of a Cowboy-Soldier.* New York: Frederick A. Stokes Co., 1931.

Schultz, Vernon B. *Southwestern Town, The Story of Willcox, Arizona.* n.p., 1964.

Sherman, James E., and Barbara H. Sherman. *Ghost Towns of Arizona.* Norman: University of Oklahoma Press, 1969.

Sonnichsen, C. L. *Billy King's Tombstone.* Tucson: University of Arizona Press, 1942.

————. *Colonel Greene and the Copper Skyrocket.* Tucson: University of Arizona Press, 1974.

————. *Tucson.* Norman: University of Oklahoma Press, 1982.

Sparks, William. *The Apache Kid, A Bear Fight, and Other True Stories of the Old West.* Los Angeles: Skelton Publishing Company, 1926.

Turner, John Kenneth. *Barbarous Mexico.* Austin: University of Texas Press, 1969.

Vanderwood, Paul J. *Disorder and Progress — Bandits, Police, and Mexican Development.* Lincoln: University of Nebraska Press, 1981.

Wachholtz, Florence, comp. *Arizona, the Grand Canyon State.* 2 vols. Westminster, CO: Western States Historical Publishers, Inc., 1975.

Wagoner, Jay J. *Arizona Territory, 1863–1912.* Tucson: University of Arizona Press, 1970.

Walker, Henry P., and Don Bufkin. *Historical Atlas of Arizona.* Norman: University of Oklahoma Press, 1979.

Walters, Lorenzo D. *Tombstone's Yesterday.* Tucson: Acme Printing Co., 1928.

Wentworth, Frank L. *Bisbee With the Big B.* Iowa City, IA: Mercer Printing Company, 1938.

Woody, Clara T., and Milton L. Schwartz. *Globe, Arizona.* Tucson: Arizona Historical Society, 1977.

## Articles

"The Arizona Rangers." *The Arizona Highway Patrolman* (Spring 1979): 11, 13.

Brayer, Herbert O. "The Cananea Incident." *New Mexico Historical Review* (October 1938): 387–415.

DeArment, Robert K. "Arizona Ranger Jeff Kidder." Tombstone *Epitaph* (National Edition, March 1981): 1, 9–11.

Donoho, Ron. "Death of an Arizona Ranger." Nevada Peace Officers Association (December 1971).

Egerton, Kearney. "The Arizona Rangers Hang Two Men." Clipping in the files of the Arizona Heritage Center, Tucson.

———. "The Arizona Rangers Settle a Labor Dispute." *Arizona Republic* (July 4, 1982).

———. "A Brazen Horse Thief's Comeuppance." *Arizona Republic* (July 12, 1981).

———. "The Case of the Cow Thief." *Arizona Republic* (March 28, 1982).

———. "Pursuit in Mexico." *Arizona Republic* (April 18, 1982).

———. "A Tale of Two Petitions." *Arizona Republic* (February 6, 1983).

Ernenwein, Leslie. "Lucky Star." *Ranch Romances* (July 8, 1949): 43–47.

Herner, Charles R. "Arizona's Cowboy Cavalry." *Arizoniana* (Winter 1964): 10–26.

Hewes, Charles. "Captain of the Arizona Rangers." *Saga* (June 1958): 53–60.

"Honor the Past . . . Mold the Future," *Gila Centennial Historical Celebration and Pageant* (1976).

Hornung, Chuck. "Fullerton's Rangers." Tombstone *Epitaph* (National Edition, January 1981): 7–9.

Jeffrey, John Mason. "Discipline in the Arizona Territorial Prison: Draconian Severity or Enlightened Administration?" *The Journal of Arizona History* (Autumn 1968): 140–154.

Jensen, Jody. "Birth of the Arizona Rangers." *Old West* (Spring 1983): 30–33.

Keen, Effie R. "Arizona's Governors." *Arizona Historical Review* (October 1930): 7–20.

Kelley, Edward J. "The Killing of Jack the Ripper." *Arizona Highways* 15 (November 1939): 20–21, 33.

Kildare, Maurice. "Arizona's Toughest Ranger." *Old West* (Summer 1967).

McCool, Grace. "With Grace McCool" (untitled articles on the Arizona Rangers). *Gateway Times* (September 14, 1961; September 16, 1961; April 16, 1964; April 19, 1964; April 23, 1964).

Myers, John Myers. "A Chivalrous Killer." *Arizona Days and Ways* (December 13, 1964).

Nichols, Roger L. "A Miniature Venice, Florence, Arizona, 1866–1910." *The Journal of Arizona History* (Winter 1975): 335–356.

Park, Joseph F. "The 1903 'Mexican Affair' at Clifton." *The Journal of Arizona History* (Summer 1977): 119–148.

Pearce, Joe. "The Killing of Arizona Rangers at the 'Battle Ground'." *Arizona Stockman* (April 1947): 7–9.

———, and Richard Summers. "Joe Pearce — Manhunter." *The Journal of Arizona History* (Autumn 1978): 249–260.

Peterson, C. O. "Naco, Arizona's Accidental Place in History." *American West* (January/February 1983): 44–47, 70–71.

Rathbun, Carl M. "Keeping the Peace Along the Mexican Border." *Harper's Weekly* 50, No. 2604 (1906).

Rattenbury, Richard. "A Portfolio of Firearms from the New Winchester Museum." *American West* (May/June 1980): 34–45.

Row, Tracy, comp. "Graphic Arts on the Arizona Frontier, The Face of Early Phoenix." *The Journal of Arizona History* (Summer 1972): 109–122.

Sharp, Patricia Maxwell. "The Maxwells of Arizona — Trackers and Lawmen." *Frontier Times* (July 1972): 34–35, 44, 46.

Shirley, Glenn. "Cap Mossman — and the Apache Devil." *True West* (November 1957): 4–6, 32–34.

Sonnichsen, C. L. "Col. W. C. Greene and Cobre Grande Copper Company." *The Journal of Arizona History* (Summer 1971): 73–100.

Spangenberger, Phil. "Thomas H. Rynning." *Guns and Ammo Guide to Guns of the Gunfighters* (1975): 31–37.

Stocher, Joseph. "The Arizona Rangers . . ." *Arizona Highways* (August 1982): 35–39.

Virgines, George E. "The Arizona Rangers . . . Birth of a Commemorative." *Guns* (June 1975): 46–47, 56–58.

———. "The Arizona Rangers." *Arms Gazette* (December 1973): 26–31, 34–35.

———. "Return of the Rangers." *Gun World Magazine* (February 1970): 82–85.

Wahmann, Russell, comp. "Front Street, Flagstaff, From Trail to Thoroughfare." *The Journal of Arizona History* (Spring 1973): 31–46.

Waltrip, Lela, and Rufus Waltrip. "Top Man of the Fearless Thirteen." *True West* (December 1970): 22–25, 72, 74.

Willson, Roscoe G. "Arizona Days and Ways, Sergeant Jeff Kidder." *Arizona Republic* (February 22, 1948).

Winsor, Mulford. "The Arizona Rangers." *Our Sheriff and Police Journal* 31, No. 6 (1936): 49–61.

# Index

Printed in the USA
CPSIA information can be obtained
at www.ICGtesting.com
LVHW010621160424
777486LV00019B/447